Physiology of Adrenocortical Secretion

Tatuzi Suzuki

Department of Physiology, Nagasaki University School of Medicine, Nagasaki, Japan

28 figures and 7 tables, 1983

S. Karger · Basel · München · Paris · London · New York · Tokyo · Sydney

Frontiers of Hormone Research

National Library of Medicine, Cataloging in Publication
Suzuki, Tatuzi
Physiology of adrenocortical secretion
Tatuzi Suzuki – Basel, New York, Karger, 1983.
(Frontiers of hormone research; v. 11)
1. Adrenal cortex – secretion 2. Pituitary-adrenal system – physiology I. Title
II. Series
W1 FR946F v. 11 [WK 750 S968p]
ISBN 3–8055–3644–5

Drug Dosage
The authors and publisher have exerted every effort to ensure that drug selection and dosage set forth in this text are in accord with current recommendations and practice at the time of publication. However, in view of ongoing research, changes in government regulations, and the constant flow of information relating to drug therapy and drug reactions, the reader is urged to check the package insert for each drug for any change in indications and dosage and for added warnings and precautions. This is particularly important when the recommended agent is a new and/or infrequently employed drug.

Contents

Preface

This monograph is designed to provide a review of the physiological aspects of adrenal cortisol and corticosterone secretion. In the opening chapters an attempt is made to present a thorough survey of research findings on the resting and stress-induced adrenocortical secretion. The subsequent chapters are a concise review of the ever-expanding field of adrenal physiology, i.e. the circadian variations of plasma corticosteroid levels and the mechanisms which control the pituitary-adrenocortical secretory activity.

I have been fortunate enough to have spent a 10-year period in the Physiological Laboratory of Dr. *Y. Satake* and Dr. *M. Wada* in Tohoku University. My interest in the field of adrenal physiology was strongly stimulated by the above two outstanding professors, to whom I should like to express my sincere gratitude. I am indebted to all my colleagues in the Physiology Department of Nagasaki University for their invaluable support. I am especially indebted to Dr. *T. Hirose* and Mrs. *K. Suzuki,* my wife, for checking the list of references, compiling the subject and author indexes and proofreading the entire manuscript, and to Dr. *H. Matsui* for preparing all illustrations. I am also indebted to Mr. *I. Matsumoto* and Miss *Y. Mine* for checking the list of references, to Miss *F. Yoshimura* and Miss *K. Kinoshita* for typing the manuscript, and to Dr. *T. Aikawa* and Dr. *S. Suzuki,* my son, for proofreading.

I am grateful to the publishers who gave me generous permission to reproduce table III from 'Nature, London' and figure 13 from the 'Japanese Journal of Physiology'. Finally, I also wish to express my appreciation to Mr. *H. Katakura* (Katakura Libri, Inc.) and the staff of S. Karger for their patient collaboration in producing this monograph.

T. Suzuki
Nagasaki, 1983

Chapter 1. Introduction

A. Indices of Adrenocortical Secretory Activity

Over a period of many years the studies of adrenal cortical secretion were mostly done by using indirect indices of adrenocortical secretory activity, such as circulating eosinophil and lymphocyte counts, adrenal ascorbic acid or cholesterol concentration, etc.

A marked decrease in eosinophil and lymphocyte counts in man and in experimental animals following exposure to stressful stimuli has been well established. However, eosinopenic and lymphopenic responses have been observed sometimes even in adrenalectomized animals. This casts a serious doubt on the specificity of circulating eosinophil (or lymphocyte) counts as an index of secretory activity of the adrenal cortex.

Ascorbic acid concentration of the rat adrenal gland was one of the most widely used indices in the earlier investigations of adrenal cortical secretion. The validity of this index, however, has been questioned because of discrepancies between the adrenal ascorbic acid and plasma corticosterone (or adrenal corticosterone secretion) responses in rats to stressful stimuli or exogenous adrenocorticotropic hormone (ACTH) [*Guillemin* et al., 1958, 1959b; *Slusher,* 1958; *Montanari and Stockham,* 1962].

Vogt [1943] was the first to use a direct method for determining adrenocortical secretion rates in experimental animals. In her pioneer experiments, the animals (dogs, cats, rabbits, goats, and pigs) were anesthetized with ether followed by chloralose (i.v.) and subjected to severe abdominal surgery for collection of adrenal venous blood samples. The samples were assayed for adrenocortical hormone by the cold exposure test of *Selye and Schenker* [1938], a method based on a prolongation of survival time in young adrenalectomized cold-exposed rats by treatment with adrenocortical hormone or adrenal venous plasma samples. A surprisingly large amount of corticosteroids was found to be continuously secreted. In the dog the mean 24-hour secretion rate of both adrenal glands per 10 kg body weight was equivalent to adrenocortical hormone content of 17,300 g adrenal gland. However, she noted the possibility that the above experimental results might be due to the special conditions, such as anesthesia and severe abdominal surgery, under which her experiments were done.

In some subsequent studies, in which the adrenal cortical secretion rates were directly evaluated, experiments were also performed under conditions of severe surgical stress [*Vogt*, 1944; *Corcoran and Page*, 1948; *Okinaka* et al., 1952; *Frank* et al., 1955]. Indeed, *Frank* et al. [1955] stated as follows:

An ideal experiment would permit sampling from an unfrightened, unanesthetized, intact animal. Since even the injection of a sedative dose of morphine may induce a fall in eosinophile count, conditions permitting the determination of the basal rate of corticosteroid secretion are not readily, if ever, attainable.

However, a method of collecting adrenal venous blood samples in conscious dogs under basal conditions was devised as early as 1927 by *Satake* et al., a method originally devised for and used in the studies of adrenal medullary secretion [*Satake*, 1955]. A modification [*Suzuki* et al., 1959a] of this method has been used in a large number of studies of adrenocortical secretion. In these studies adrenal venous blood samples were analyzed for 17-hydroxycorticosteroids (17-OHCS) by the method of *Nelson and Samuels* [1952].

Adrenal vein cannulation and collection of adrenal venous blood samples were conducted in the following ways. In order to collect adrenal venous blood samples from conscious dogs without provoking any pain, the lumbar areas of the animal were previously deafferented, i.e. the dorsal spinal roots from T_{11} to L_3 were cut under pentobarbital anesthesia. 3 weeks or more were allowed to elapse for a recovery from surgery. On the day before actual experiment the lumboadrenal vein was exposed through the lumbar route and its small branches were tied and cut off between two ligatures. Then a loose loop of a long silk thread was passed round the lumboadrenal vein between the inferior vena cava and the adrenal gland. A cannula attached to a short rubber tube was introduced into the lumboadrenal vein just lateral to the adrenal gland and was fixed. The cannula and rubber tube were filled with heparin-Ringer solution and the rubber tube was clamped. Actual experiment was started usually 18 or more hours after the completion of the adrenal vein cannulation. At the time of adrenal venous blood sample collection the clamp laid on the rubber tube was removed and the long silk thread was gently pulled so as to direct the blood flow from the adrenal gland toward the exterior through the glass cannula and rubber tube. At the end of the period of blood sample collection the silk thread round the adrenal vein was released, the cannula and rubber tube were filled again with heparin-Ringer solution, and the rubber tube was clamped.

Other methods for collecting adrenal venous blood samples from conscious dogs were independently devised by *Hume and Nelson* [1955a] and by *Endröczi* et al. [1958]. The method of the former investigators has widely been used in experiments on conscious or anesthetized dogs.

A method of direct determination of adrenal cortical secretion rates in conscious sheep with an adrenal gland autotransplanted to a carotid artery-jugular vein loop [*McDonald* et al., 1958] has been used by Australian physiologists. Arterial perfusion of the adrenal gland [*Vogt*, 1951a; *Hilton* et al., 1958; *Redmond and Bell*, 1974; *Aikawa* et al., 1981b], in vitro superfusion of the adrenal glands [*Tait* et al., 1967], and in vitro incubation of dispersed adrenal cells [*Kloppenborg* et al., 1968] are useful methods for evaluating direct effect of various substances on adrenocortical secretion. For determination of hourly or daily secretion rate of adrenocortical hormone, particularly in man, the isotope dilution method [*Peterson*, 1959] has been used.

Peripheral plasma corticosteroid concentration has been and is still at present one of the most widely used measures of adrenocortical secretory activity. It is a useful but not very reliable index, since it can be modified by several factors other than the adrenocortical secretion rate, such as the volume of corticosteroid distribution, the rates of metabolic elimination and excretion of corticosteroids, etc.

Urinary excretion rates of corticosteroids, 17-ketosteroids and 17-ketogenic steroids were also used as indices of adrenocortical secretory activity. The reliabilities of these indices are questioned.

B. Resting Adrenocortical Secretion Rate

In a number of studies of adrenocortical secretion, *Suzuki* et al. [1959a,b, 1960, 1962, 1963, 1964a,b, 1965a,b, 1966a,b, 1967, 1968, 1971, 1972, 1973, 1975a] determined the resting adrenal 17-OHCS secretion rate in non-stressed conscious dogs using a modification of the method of *Satake* et al. [1927] for collections of adrenal venous blood samples. In 113 conscious dogs the mean value of resting secretion rates of 17-OHCS with its standard error was 0.10 ± 0.0046 µg/kg/min

Table I. Resting adrenocortical secretion rate in conscious dogs determined by the method of *Hume and Nelson* [1955a]

References	Number of dogs	Number of determina- tions	Adrenal 17-OHCS secretion rate, μg/min	
			mean ± SEM[1]	SD[2]
Egdahl and Richards [1956a]	18	35	2.3 ± 0.4	2.6
Egdahl and Richards [1956b]	10	56	2.3	
Egdahl et al. [1956]	5	13	0.38 ± 0.12	0.45
Nelson et al. [1956]	16	99	2.8 ± 0.6	
Egdahl [1959]	7	7	2.4 ± 0.9	2.4
Kwaan and Bartelstone [1959]		22	2.2 ± 0.2	
Zukoski [1966]	2	2	2.1	
Roy [1969]	20	20	2.2 ± 0.1	0.6

[1] Standard error of the mean.
[2] Standard deviation.

(micrograms per kilogram of body weight per minute; number of determinations = 225, standard deviation = 0.069 μg/kg/min). Utilizing the same method, *Marotta* et al. [1963] also determined the resting 17-OHCS secretion rates in the dog. They reported that the mean value with its standard error in 8 conscious dogs was 0.09 ± 0.02 μg/kg/min.

Some investigators have evaluated the resting adrenocortical secretion rates in non-stressed conscious dogs by collecting adrenal venous blood samples by means of the method of *Hume and Nelson* [1955a]. Data are presented in table I. In their reports the adrenal 17-OHCS secretion rates were expressed as μg/min rather than μg/kg/min.

Chapter 2. Adrenal Cortical Secretion in Response to a Variety of Internal and External Environmental Stimuli

A. Ether Anesthesia

A significant adrenal ascorbic acid depletion in rats in response to ether anesthesia has been observed [*Royce and Sayers*, 1958a; *Kitay* et al., 1959a, b; *Mitamura*, 1960a; *Giuliani* et al., 1961; *Montanari and Stockham*, 1962; *Barrett and Stockham*, 1965], a response completely abolished by hypophysectomy [*Mitamura*, 1960a].

Beigelman et al. [1956] studied the effect of anesthetics on adrenal corticosteroid secretion in rats. They performed the left renal vein cannulation under ether or pentobarbital anesthesia for collecting adrenal venous blood samples. During the first 10-min period of sample collection, the rate of adrenal corticosterone secretion was found to be markedly influenced by the type of anesthetic used in experiments; the corticosterone secretion rate in ether-anesthetized rats was significantly higher than that in pentobarbital-anesthetized rats. They stated that the elevated corticosterone secretion observed in ether-anesthetized rats might partly be due to the stimulatory effect of ether anesthesia on pituitary-adrenocortical secretory activity. However, it was not possible in their experiments to separate the effect on adrenal corticosterone secretion of ether anesthesia from that of surgical procedures. *Nelson* et al. [1956] evaluated adrenal 17-OHCS secretion in dogs whose adrenal vein was cannulated under ether or pentobarbital anesthesia. On the day of adrenal vein cannulation the mean 17-OHCS secretion rate in ether-anesthetized dogs was 10.3 ± 0.7 (SEM) µg/min (unless otherwise specified mean values are given in subsequent pages with their standard errors) and that in pentobarbital-anesthetized dogs was 6.1 ± 0.8 µg/min. The above experimental results might suggest that ether anesthesia is more potent in stimulating adrenocortical secretion than pentobarbital anesthesia. Similar findings were also reported by *Endröczi* et al. [1958].

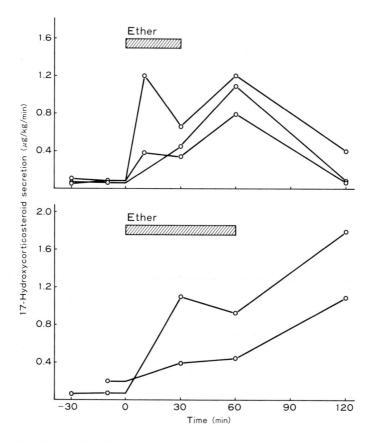

Fig. 1. Elevation of adrenal 17-OHCS secretion rate in dogs during and after 30- and 60-min periods of ether inhalation. [Based on data from *Suzuki* et al., 1959a.]

Suzuki et al. [1959a] have attempted to evaluate directly adrenocortical secretion in dogs in response to ether anesthesia without surgical stress. In their study the adrenal vein was cannulated without anesthesia and experiments were started about 18 h after completion of the cannulation. Ether was inhaled for 30–60 min. During ether inhalation the adrenal 17-OHCS secretion rate increased from the basal levels of 0.06–0.20 to 0.38–1.2 μg/kg/min (fig. 1). At 30–60 min after the end of ether inhalation period the secretion rate further increased to 0.79–1.8 μg/kg/min. In the experiments of *Egdahl* [1965/66] a marked increase in adrenal 17-OHCS secretion in dogs was observed between 15 and

120 min following the start of ether inhalation. He stated that ether was an extremely potent, but variable, stimulus to pituitary-adrenal secretion. A marked increase in adrenal 17-OHCS secretion in dogs following ether administration was also observed by *Roy* [1969]; the mean adrenal 17-OHCS secretion rate before ether anesthesia was 2.31 ± 0.19 µg/min and that 30 and 60 min after the onset of ether inhalation was 10.61 ± 0.66 and 13.06 ± 0.78 µg/min, respectively.

A marked elevation of plasma corticosterone levels in rats after exposure to ether vapor has been reported by a great number of investigators [*Montanari and Stockham, 1962; Kraicer and Logothetopoulos, 1963a; Slusher, 1964; Barrett and Stockham, 1965; Cann* et al., 1965; *Arimura* et al., 1967; *Zimmermann and Critchlow, 1967; Feldman* et al., 1968; *Greer and Rockie, 1968; Grimm and Kendall, 1968; Voloschin* et al., 1968; *Palka* et al., 1969; *Cook* et al., 1973; *Van Delft* et al., 1973; *Wilson and Critchlow, 1973/74; Allen and Allen, 1974; Feldman and Conforti, 1976a; Jobin* et al., 1976; *Sakakura* et al., 1976b; *Yasuda and Greer, 1976b, 1978; Yasuda* et al., 1976; *Tang and Phillips, 1977; Sithichoke and Marotta, 1978; Makara* et al., 1980; *McMurtry and Wexler,* 1981; *Vernikos* et al., 1982]. A rise in plasma corticosterone levels in rats following etherization is completely abolished by hypophysectomy [*Yasuda and Greer*, 1978]. *Gibson* et al. [1979] showed an ether-induced elevation of plasma corticosteroids in mice.

A significant increase in adrenal corticosterone concentrations in rats [*Barrett and Stockham, 1965; Zimmermann and Critchlow,* 1967] and that in in vitro production of corticosteroids of excised adrenal glands in rats [*Van Delft* et al., 1973] following exposure to ether vapor were also reported.

The site of action of ether on pituitary-adrenocortical activity was suggested by *Matsuda* et al. [1964] to be localized in the median eminence and by *Makara* et al. [1980] in some brain structures outside the medial basal hypothalamus.

B. Barbiturate Anesthesia

In rats pentobarbital anesthesia has no adrenal ascorbic acid depleting effect [*Long,* 1947; *Van Peenen and Way,* 1957; *Mitamura,* 1960a, b], whereas hexobarbital anesthesia induces a small depletion of adrenal ascorbic acid [*Mitamura,* 1960a].

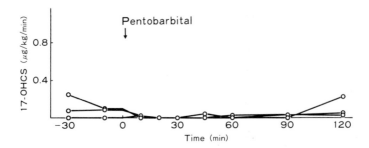

Fig. 2. Effect of pentobarbital anesthesia (i.v. injection of 25 mg/kg sodium pentobarbital) on adrenal 17-OHCS secretion in dogs. [Based on data from *Suzuki* et al., 1962.]

Walker et al. [1959b] failed to find any significant change in adrenal 17-OHCS secretion in the dog following i.v. injection of thiopen-
tal (11.8 mg/kg). Their observation was made, however, in only 1 dog.
Suzuki et al. [1962] examined in 6 dogs the effect of barbiturate anesthesia on adrenocortical secretion. Non-stressed conscious dogs were i.v.
injected with sodium pentobarbital (25 mg/kg) and were deeply
anesthetized. During the period of anesthesia adrenal 17-OHCS secretion decreased from the basal secretion rates of 0.09 ± 0.03 µg/kg/min
to the levels below 0.02 ± 0.002 µg/kg/min (fig. 2). In their studies a
significant depression of adrenal 17-OHCS secretion in dogs following
i.v. injection of sodium hexobarbital (50 mg/kg) was also observed.
Roy [1969] also demonstrated a significant decrease in adrenal
17-OHCS secretion rate in dogs during pentobarbital anesthesia.

Egdahl [1961b] demonstrated that a markedly elevated adrenal
17-OHCS secretion in decorticated dogs was promptly depressed by
i.v. administration of pentobarbital.

Rerup and Hedner [1962] showed a significant decrease in plasma
corticosterone in rats during pentobarbital anesthesia. *Yasuda* et al.
[1976] observed a significant depression of plasma corticosterone levels
in rats in the p.m. but not in the a.m. following intraperitoneal (i.p.) administration of pentobarbital. By contrast, the rise of adrenal [*Pincus
and Hirai*, 1964; *Barrett and Stockham*, 1965] and plasma [*Barrett and
Stockham*, 1965] corticosterone levels in pentobarbital-anesthetized
rats has also been reported. In man [*Siker* et al., 1956] and in monkeys
[*Harwood and Mason*, 1957] pentobarbital anesthesia depresses the peripheral plasma levels of 17-OHCS. In cats the circadian elevation of

plasma 17-OHCS levels is completely abolished by i.p. administration of pentobarbital [*Krieger and Krieger*, 1967; *Krieger* et al., 1968].

The effect of barbiturate anesthesia on the adrenocortical secretory response to stressful stimuli has been examined in numerous studies. It has been shown that pentobarbital anesthesia suppresses the adreno-cortical response to some but not to other types of stimuli.

Adrenal ascorbic acid depletion in rats in response to morphine [*Munson and Briggs*, 1955; *Ohler and Sevy*, 1956; *Van Peenen and Way, 1957; Mitamura*, 1960b] and aspirin [*Van Peenen and Way*, 1957], but not to histamine [*Van Peenen and Way*, 1957; *Mitamura*, 1960b; *Leeman* et al., 1962], epinephrine [*Van Peenen and Way*, 1957; *Mitamura*, 1960b] and insulin [*Mitamura*, 1960b], is suppressed by pretreatment with pentobarbital. Adrenal ascorbic acid depletion in rats induced by centrifugal stress (vertical spinning) is inhibited by pentobarbital, hexo-barbital and phenobarbital anesthesia [*Pohujani* et al., 1969].

In the dog, the adrenal 17-OHCS secretory response to ether stress [*Egdahl*, 1965/66], atropine [*Otsuka*, 1966] and immobilization [*Suzuki* et al., 1968] is markedly suppressed or completely abolished, and that to eserine [*Otsuka*, 1966] and pilocarpine [*Suzuki* et al., 1975a] is par-tially impaired by pentobarbital anesthesia. In contrast, adrenal 17-OHCS secretion in dogs in response to hypoxic hypoxia [*Marotta* et al., 1963], tetramethylammonium [*Otsuka*, 1966], anaphylactic shock [*Suzuki* et al., 1966a], histamine [*Tanigawa*, 1967] and nicotine [*Suzuki* et al., 1973] is not suppressed by pentobarbital anesthesia.

Greer and Rockie [1968] observed that an ether-induced elevation of plasma corticosterone levels in rats was markedly suppressed by pentobarbital anesthesia.

C. Morphine

A significant depletion of adrenal ascorbic acid in rats following systemic administration of morphine has been presented [*Nasmyth*, 1954; *Briggs and Munson*, 1955; *George and Way*, 1955a, b; *Munson and Briggs*, 1955; *Van Peenen and Way*, 1957; *Mitamura*, 1960a, b; *Nikodijevic and Maickel*, 1967], a response completely abolished either by hypophysectomy [*George and Way*, 1955a, b; *Mitamura*, 1960a] or by pretreatment with pentobarbital [*Munson and Briggs*, 1955; *Ohler and Sevy*, 1956; *Mitamura*, 1960b].

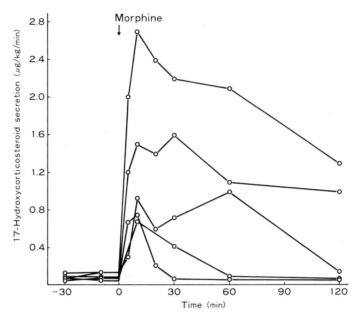

Fig. 3. Adrenal 17-OHCS secretion in dogs in response to morphine (i.v. injection of 8 mg/kg morphine hydrochloride). [Based on data from *Suzuki* et al., 1959b.]

Suzuki et al. [1959b] were the first directly to evaluate the effect of morphine on adrenal cortical secretion in the dog. 5 conscious dogs were injected i.v. with 8 mg/kg of morphine. As shown in figure 3, the adrenal 17-OHCS secretion rates were elevated from the resting levels of 0.05–0.13 µg/kg/min to the peak levels of 0.67–2.7 µg/kg/min, which were attained in 3 out of 5 dogs at 10 min, in 1 dog at 30 min and in another dog at 60 min after the injection of morphine. The duration of increased adrenocortical secretion was 20–30 min in 2 dogs, 60 min in 1 dog and 120 min or longer in 2 dogs.

Slusher and Browning [1961], *Nikodijevic and Maickel* [1967] and *Lotti* et al. [1969] demonstrated a significant elevation of plasma corticosterone levels in rats systemically injected with morphine. Confirmatory observations in rats were reported by *Kuwamura* [1972] and *Kuwamura* et al. [1973]. These investigators and *Hirai* et al. [1970] also observed a marked increase in adrenal corticosterone concentration in rats after systemic administration of morphine. In rats an i.p. injection of morphine induces a significant increase in corticosteroid production in vitro of excised adrenal glands [*Van Ree* et al., 1976b]. Plasma corticosterone levels in mice [*Gibson* et al., 1979] and adrenal corticosteroid levels in cats [*Borrell* et al., 1974] have been shown to be elevated following i.p. injection of morphine.

In the study of the site of stimulatory action of morphine on adrenocortical secretion, *Lotti* et al. [1969] demonstrated that the adrenal ascorbic acid depletion and elevation of plasma corticosterone levels in rats were induced by injection of a small amount of morphine into the mid-hypothalamic area, involving either the whole or part of the paraventricular, anterior, ventro- and dorsomedial hypothalamic nuclei. The site of the stimulatory action of morphine was also studied by *Van Ree* et al. [1976a]. Microinjection of morphine into the medial and ventral part of the mid-hypothalamus in rats was found to produce a significant elevation in plasma corticosterone levels and a significant increase in in vitro production of corticosteroids of excised adrenal glands. The most effective microinjection of morphine in elevating pituitary-adrenocortical secretory activity was that in or adjacent to the rostral part of the arcuate nucleus. The findings suggested that morphine might stimulate the neurosecretory cells in the arcuate nucleus to release ACTH-releasing factor (CRF).

Suggestive inhibitory action of morphine on evening circadian peak of plasma corticosterone levels in rats was noted by *Slusher and Browning* [1961]; plasma corticosterone levels at 5:00 p.m. in rats injected with morphine (10 mg/kg, i.v.) 8 h previously were 15.3 ± 1.7 µg/100 ml, whereas those in control rats injected with saline were 31.5 ± 1.1 µg/100 ml.

Inhibition by morphine of pituitary-adrenocortical secretory response to stressful stimuli has been demonstrated by a number of investigators. In morphine-tolerant rats, which had been injected daily with morphine for 4 days prior to the observation, the adrenal ascorbic acid depletion in response to histamine [*Briggs and Munson*, 1955; *Munson and Briggs*, 1955; *Wexler*, 1963] and surgical stress but not to bacterial polysaccharide [*Wexler*, 1963] is abolished by pretreatment with morphine. *Nikodijevic and Maickel* [1967] observed that the adrenal ascorbic acid and plasma corticosterone responses in rats to a 2-hour exposure to cold (4 °C) were blocked by pretreatment with 3 doses of morphine (30 mg/kg each during 24 h). In pentobarbital-anesthetized rats, the adrenal ascorbic acid depletion in response to

epinephrine [*Briggs and Munson*, 1955; *Ohler and Sevy*, 1956], histamine [*Briggs and Munson*, 1955; *Munson and Briggs*, 1955; *Leeman* et
al., 1962], vasopressin or Pitressin [*Briggs and Munson*, 1955; *Ohler and
Sevy*, 1956; *De Wied* et al., 1958; *Leeman* et al., 1962], serotonin [*Leeman* et al., 1962], dopamine [*King and Thomas*, 1968], prostaglandin E_1
[*Peng* et al., 1970], surgical stress [*Briggs and Munson*, 1955; *Munson
and Briggs*, 1955; *Ohler and Sevy*, 1956; *McCann*, 1957], whole body
X-ray irradiation [*Bacq and Fischer*, 1957] and the rise of plasma corticosterone levels in response to histamine, Pitressin and surgical stress
[*Leeman* et al., 1962] have been found to be depressed or completely
abolished by pretreatment with morphine.

Inhibition by morphine of stress-induced adrenocortical secretion
is not due to the direct inhibitory action of morphine at the adrenal
level, since morphine does not impair the secretory responsiveness of
the adrenal cortex to ACTH [*Briggs and Munson*, 1955; *Munson and
Briggs*, 1955; *McCann*, 1957; *Wexler*, 1963; *Kuwamura*, 1972]. The findings that in pentobarbital-anesthetized rats pretreatment with morphine depresses the pituitary-adrenocortical secretory response to histamine, vasopressin (or Pitressin), serotonin and surgical stress but not to
hypothalamic extract containing CRF [*Leeman* et al., 1962] suggest that
morphine might exert its inhibitory effect by the action on the hypothalamus or extrahypothalamic brain structures but not anterior pituitary.

D. Ethanol

A fall in adrenal ascorbic acid content in response to ethanol has
been observed in rats by *Smith* [1950, 1951], *Forbes and Duncan* [1951,
1953], *Řežábek* [1957] and *Czaja and Kalant* [1961], and in guinea pigs
by *Forbes and Duncan* [1951, 1953]. Adrenal cholesterol depletion in
rats and guinea pigs following administration of ethanol has also been
reported [*Smith*, 1950, 1951; *Forbes and Duncan*, 1951, 1953; *Czaja and
Kalant*, 1961]. The above adrenal responses to ethanol in rats or guinea
pigs have been shown to be abolished by hypophysectomy [*Smith*,
1950, 1951; *Forbes and Duncan*, 1951; *Řežábek*, 1957]. *Kalant* et al.
[1963] showed an elevation of corticosteroid production in vitro of excised adrenal glands of ethanol-treated rats.

Suzuki et al. [1972] have performed direct evaluation of adrenocortical secretion in dogs in response to ethanol, the data of which are pre-

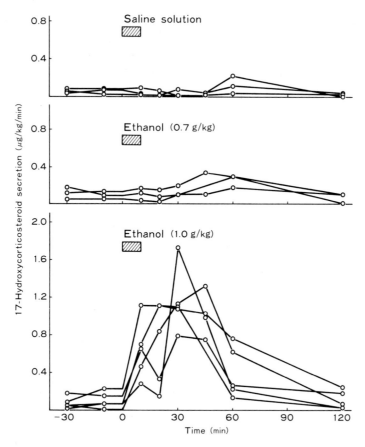

Fig. 4. Adrenocortical secretory response in unanesthetized dogs to ethanol (i.v. infusion of 0.7 and 1.0 g/kg ethanol over 10 min). [Based on data from *Suzuki* et al., 1972.]

sented in figure 4. In 3 dogs infused i.v. with 0.7 g/kg of ethanol over 10 min, the blood ethanol concentration at the end of ethanol infusion was 95 ± 14 mg/100 ml. The animals became quiet and depressive but did not sleep during and after the infusion of ethanol. In these 3 dogs either slight increase or no change in adrenal 17-OHCS secretion was observed after ethanol infusion. 5 dogs infused i.v. with 1.0 g/kg of ethanol over 10 min lay down and deeply slept for 10–25 min. The blood ethanol concentration just after the end of ethanol infusion was 206 ± 9 mg/100 ml and adrenal 17-OHCS secretion markedly increased without exception. It increased from the resting secretion rates of 0.09 ± 0.023 to 1.21 ± 0.15 µg/kg/min. Control experiments were performed in 3 conscious dogs. Adrenal 17-OHCS secretion was observed to be unchanged following 10-min i.v. infusion of 0.9% saline solution.

An elevation of plasma corticosterone levels in rats [*Ellis*, 1966] and in mice [*Kakihana* et al., 1968, 1971] following ethanol administration has been reported. *Ellis* [1966] noted that a rise in plasma corticosterone levels in rats in response to ethanol was partially blocked by pretreatment with pentobarbital and completely abolished either by pentobarbital plus morphine or by hypophysectomy.

Linkola et al. [1979] observed in man an initial depression and a subsequent elevation of peripheral plasma cortisol levels following ingestion of ethanol.

E. Immobilization

Significant depletion of adrenal ascorbic acid in rats following a 90-min immobilization with adhesive tape has been reported [*Guillemin*, 1955].

Knigge et al. [1959] evaluated adrenal corticosterone secretion in rats following immobilization stress (binding the paws together by adhesive tape and wrapping the animals in a towel). Corticosterone secretion rates were measured in their study by adrenal vein cannulation and collection of adrenal venous blood samples under pentobarbital anesthesia. The basal secretion rate of corticosterone was 80 ± 6 (SD) µg/100 mg adrenal/h. At 1 and 2 h after immobilization the secretion rate increased to 192 ± 13 and 180 ± 13 µg/100 mg/h. Then it decreased to 20 ± 3 µg/100 mg/h 4 h after the stress and returned to preimmobilization level at 12 h.

Walker et al. [1959b] studied the effect of immobilization stress on adrenocortical secretion in the dog. In their study, 2 conscious dogs were immobilized by strapping to a table for 100–270 min. In 1 dog which was quiet during the whole period of immobilization, the adrenal blood flow gradually decreased from the basal level of 3.8 to 1.4 ml/min and an increase in adrenal 17-OHCS secretion was observed when adrenal blood flow was low; the maximal adrenal 17-OHCS secretion rate following immobilization was 3.8 µg/min, which was at-

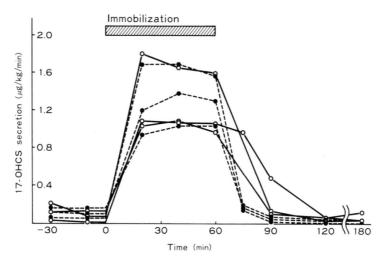

Fig. 5. Time course of adrenocortical secretory response in conscious dogs to a 60-min immobilization in the supine (O) or prone (●) position. Experimental results in 6 out of 9 dogs are illustrated in the figure (see the text). [Based on data from *Suzuki* et al., 1968.]

tained at 210 min after the start of immobilization. In the other dog in which immobilization was resisted, adrenal 17-OHCS secretion was increased from 7.3 to 11.2 µg/min, an increase mainly due to a rise in adrenal blood flow.

Suzuki et al. [1968] also performed direct evaluation of adrenocortical secretion in dogs in response to immobilization stress. 9 conscious and 5 anesthetized dogs were immobilized by strapping in the prone or supine position on a table for 1 h. 6 out of 9 conscious dogs began to violently struggle and snarl soon after the start of immobilization. These signs of excitement persisted during the whole course of immobilization. A marked increase in adrenal 17-OHCS secretion was found without exception at 20, 40 and 60 min after the start of immobilization. The basal rates and peak values during immobilization were 0.02–0.21 and 1.05–1.80 µg/kg/min, respectively. Soon after being freed from immobilization the animals lay down quietly and the 17-OHCS secretion rate fell down to the resting values by 15–30 min in 5 dogs and 60 min in 1 dog. The time courses of adrenal 17-OHCS secretory response to immobilization stress in these 6 conscious dogs are presented in figure 5. 1 of 3 other conscious dogs was quiet during the

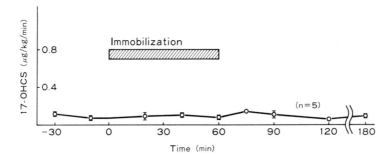

Fig. 6. Effect of a 60-min immobilization on adrenal 17-OHCS secretion rate (mean ± SEM) in 5 pentobarbital-anesthetized dogs. [Based on data from *Suzuki* et al., 1968.]

whole course of observation. The adrenal 17-OHCS secretion rate varied during and after the immobilization period from 0.02 to 0.27 µg/kg/min, which was within the range of physiological variation of resting secretion rates. In another conscious dog no signs of excitement were observed during the first two thirds of the immobilization period. A slight excitement occurred at 43 min and the animal began to struggle and snarl 2 min later. Adrenal 17-OHCS secretion, which showed no change at 20 and 40 min, was found to be markedly increased at 60 min after the start of immobilization. In summary, adrenal 17-OHCS secretion in conscious dogs was found to be consistently increased by immobilization stress, when immobilization was resisted. In 5 pentobarbital-anesthetized dogs no significant change in adrenal 17-OHCS secretion was observed during and after a 1-hour immobilization (fig. 6).

Ganong et al. [1955b] examined the effect of immobilization stress on peripheral plasma concentration of 17-OHCS in dogs. The animals were immobilized in a cage resembling a Pavlov stand for 2 h. Plasma 17-OHCS levels were elevated from the basal values of 4.4 ± 1.5 to 26.3 ± 5.2 µg/100 ml after a 2-hour immobilization. This response was observed independently whether or not the animals struggled during the immobilization period and whether or not the animals had been repeatedly handled. *Betz and Ganong* [1963] also reported a significant elevation of plasma 17-OHCS levels in dogs following immobilization.

A significant rise in plasma corticosterone levels in the rat [*Knigge*, 1961; *Purves and Sirett*, 1967; *Zimmermann and Critchlow*, 1967; *Palka*

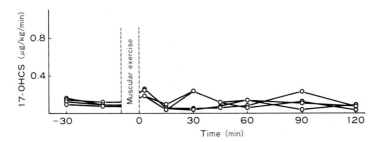

Fig. 7. Effect of light muscular exercise on adrenal 17-OHCS secretion in dogs.
[Based on data from *Suzuki* et al., 1967.]

et al., 1969; *Feldman* et al., 1970; *Dunn* et al., 1972; *Dallman and Jones*,
1973; *Sakakura* et al., 1976a, b; *Sun* et al., 1979; *McMurtry and Wexler*,
1981; *De Souza and Van Loon*, 1982] and in the mouse [*Barlow* et al.,
1975] in response to immobilization stress has been observed.

F. Muscular Exercise

A significant depletion of adrenal ascorbic acid in rats [*Ratsima-manga*, 1939; *Božović and Kostial-Živanović*, 1952; *Karaulova*, 1973]
and in guinea pigs [*Ratsimamanga*, 1939] and that of adrenal choles-terol in guinea pigs [*Knouff* et al., 1941] following muscular exercise
have been observed.

Suzuki et al. [1958] performed direct evaluation of adrenal cortical
secretion in the dog in response to muscular exercise. 5 conscious dogs,
paced by an attendant on a bicycle, were run 5.1–14.7 km. In 4 dogs a
marked increase in adrenal 17-OHCS secretion was observed after ex-haustive exercise. In 1 dog, however, which showed no sign of fatigue
during and after running, only a slight increase in adrenal 17-OHCS
secretion was observed. Adrenal 17-OHCS secretion in dogs in re-sponse to non-exhaustive and exhaustive muscular exercise was reex-amined by *Suzuki* et al. [1967]. Experiments were done on 12 conscious
dogs which were run 1.2–13.1 km. In 4 dogs in which no sign of fatigue
was observed during and after running, there was no definite change in
adrenal 17-OHCS secretion rate (fig. 7). In 4 other dogs, which seemed
to be somewhat tired but were not completely exhausted, the secretion
rate increased only transitorily. The secretion rates before running

Table II. Effect of exhausting muscular exercise on adrenocortical secretion in the dog[1]

Dog No.	Adrenal 17-hydroxycorticosteroid secretion rate, µg/kg/min								
	minutes before		minutes after end of running						
	30	10	3	15	30	45	60	90	120
1	0.06	0.03	1.49	1.12		0.74	0.29	0.06	
2	0.23	0.22	2.14	2.00	0.18	0.14	0.84	0.37	0.12
3	0.17	0.23	1.97	1.46	1.65	1.45	1.68	2.02	1.53
4	0.03	0.18	0.06	0.54	1.11	0.80	0.75	0.73	0.88

[1] Data from *Suzuki* et al. [1967].

were 0.04–0.30 µg/kg/min and the maximum secretion rates after running were 0.49–0.84 µg/kg/min. In 4 completely exhausted dogs vigorous panting and salivation were observed during running. At the end of running the animals fell down, streched their legs and refused to run further. Marked rises in the rate of heart beat and body temperature were observed. The adrenal 17-OHCS secretion rates were elevated from the resting levels of 0.03–0.23 µg/kg/min to the maximum levels of 1.11–2.14 µg/kg/min after exhaustive muscular exercise (table II).

Foss et al. [1971] studied the effect of light, heavy and exhaustive muscular exercise in dogs on peripheral plasma 11-OHCS levels. During light muscular exercise only inconsistent changes in plasma 11-OHCS levels were observed. However, during heavy exercise the plasma 11-OHCS levels were elevated from the resting value of 1.9 ± 0.4 to 5.3 ± 1.5 µg/100 ml and were further elevated at exhaustion to 8.0 ± 1.7 µg/100 ml.

Plasma concentration of cortisol in guinea pigs [*Viru and Äkke*, 1969] and that of corticosterone in rats [*Chin and Evonuk*, 1971] are significantly depressed by exhaustive but unaffected by non-exhaustive muscular exercise. However, *Craig and Griffith* [1976] failed to find significant changes in plasma corticosterone levels in rats following muscular exercise. In contrast, *Frenkl* et al. [1975] and *Viru* [1976] reported a significant elevation of plasma corticosterone levels in rats after muscular exercise. *James* et al. [1970] found a small increase in peripheral plasma cortisol concentrations in the horse after muscular exercise.

Studies by a number of investigators of the effect of muscular exercise in man on peripheral plasma corticosteroid levels have revealed rather conflicting results. A rise in plasma concentrations of 17-OHCS [*Crabbé* et al., 1956; *Nazar*, 1965], 11-OHCS [*Wenzkat* et al., 1968] and cortisol [*Lehnert* et al., 1968; *Follenius and Brandenberger*, 1974; *Gawel* et al., 1979] in response to muscular exercise has been reported. It has also been reported that the light muscular exercise produces inconsistent results, whereas the heavy exercise tends to cause an elevation of plasma levels of 17-OHCS [*Nazar*, 1966] and cortisol [*Few*, 1974]. *Davies and Few* [1976] noted that peripheral plasma cortisol levels were raised during and after muscular exercise under hypoxic but not normoxic condition. *Lewis* [1957] failed to find changes in plasma levels of cortisol and corticosterone after muscular exercise. *Cornil* et al. [1965] and *Raymond* et al. [1972] observed a significant fall in plasma cortisol levels following muscular exercise. *Froesch* et al. [1954], *Kägi* [1955] and *Staehelin* et al. [1955] noted that a rise in plasma 17-OHCS concentrations followed by a fall far below the resting levels was produced by muscular exercise. By contrast, *Sutton* et al. [1969] observed an initial fall in plasma cortisol levels during muscular exercise which was followed by a rise.

Few [1974] studied the effect of muscular exercise on the plasma cortisol levels and the specific activity of plasma cortisol in man injected i.v. with ^{3}H-cortisol. During and after heavy muscular exercise there occurred a marked fall in specific activity of plasma cortisol and a significant rise in plasma cortisol levels, thus indicating that the latter is due to an increase in adrenal cortical secretion. During and after light exercise, however, the specific activity continued to fall slowly and plasma cortisol levels fell only insignificantly, thus suggesting the latter is largely due to an increase in the rate of removal of plasma cortisol.

G. Hypoxia

Adrenal ascorbic acid depletion in rats following exposure to low atmospheric pressure for 2 h has been observed [*Aschan*, 1953].

Marotta et al. [1963] determined the adrenal 17-OHCS secretion rates in conscious and anesthetized dogs decompressed in a continuously ventilated decompression chamber to a simulated altitude of

17,000 feet (about 5,100 m) for 2 h; O_2 tension in the decompression chamber decreased from the pre-decompression level of 150 to 78 mm Hg. In pentobarbital-anesthetized dogs the adrenal 17-OHCS secretion rates increased from the resting levels of 0.06 ± 0.01 to 0.32 ± 0.10 µg/kg/min within 30 min after the start of atmospheric decompression and attained the maximal levels (7-fold increase) at 1.5 h. A significant (4- to 8-fold) increase in adrenal 17-OHCS secretion during exposure to low atmospheric pressure was also observed in conscious dogs. In control experiments in which conscious and anesthetized dogs respired ambient air in a decompression chamber at ground level, no significant change in adrenal 17-OHCS secretion occurred.

Subsequently, an elevation in adrenal secretion of 17-OHCS [*Hirai* et al., 1963; *Marotta* et al., 1965], 11-OHCS [*Marotta*, 1972b] and cortisol [*Marotta* et al., 1973] in pentobarbital-anesthetized dogs following induction of hypoxia by breathing a gas mixture of 10% O_2 in N_2 has been reported. Bilateral denervation of the aortic bodies by sectioning the vagi in the cervical region and that of the carotid bodies at the carotid bifurcation abolish the adrenal secretion of 17-OHCS [*Lau and Marotta*, 1969] and 11-OHCS [*Marotta*, 1972a] in pentobarbital-anesthetized dogs in response to hypoxic hypoxia (breathing 10% O_2 in N_2), suggesting the participation of the aortic and carotid bodies in adrenocortical secretory response to hypoxia. *Marotta* [1972a] noted that in dogs with either denervated aortic or carotid bodies the maximal adrenal 11-OHCS secretory response to hypoxia was 30–40% of that in intact dogs and suggested that both aortic and carotid bodies equally participate in the adrenocortical secretory response to hypoxia. Hypoxia induces a marked elevation of adrenal cortisol and corticosterone secretion rates in conscious calves [*Bloom* et al., 1976].

Marotta et al. [1976] examined the effect of hypoxic hypoxia on plasma corticosterone levels in rats. Hypoxia was induced by allowing the animals to breathe a gas mixture of 10% O_2 in N_2 for 1 h in a chamber with ventilating system. The mean plasma concentration of corticosterone in the control group was 4.3 ± 0.5 µg/100 ml and it was elevated in the hypoxic group to 33.6 ± 3.6 µg/100 ml. *Boddy* et al. [1974] observed an increase in plasma corticosteroid concentration in maternal but not in fetal sheep during hypoxic period (breathing of a gas mixture of 9% O_2 and 3% CO_2 in N_2). However, *Biddulph* et al. [1959] failed to find in pentobarbital-anesthetized dogs any significant changes in plasma concentrations of 17-OHCS and corticosterone-like

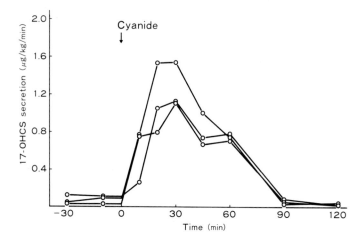

Fig. 8. Adrenal 17-OHCS secretion rates in conscious dogs before and after s.c. injection of 1.5 mg/kg potassium cyanide. [Based on data from *Suzuki* et al., 1965b.]

fraction and in adrenal ascorbic acid concentration following exposure to hypoxia induced by breathing 5% O_2 in N_2.

Adrenal ascorbic acid depletion in rats in response to cyanide hypoxia (i.p. injection of 0.4–3.0 mg/kg potassium cyanide) has been noted [*Anichkov* et al., 1960].

Suzuki et al. [1965b] studied the effect of cyanide hypoxia on adrenal 17-OHCS secretion in conscious dogs. In 3 dogs injected s.c. with 1.5 mg/kg potassium cyanide, the adrenal 17-OHCS secretion rate increased within 10 min after the injection from the resting levels of 0.03–0.12 µg/kg/min and attained the peak levels (1.11–1.55 µg/kg/min) at 30 min (fig. 8). Thereafter, it decreased and returned to the resting levels by 90 min. Similar results were obtained in 2 dogs injected s.c. with 2 or 3 mg/kg of cyanide. The adrenal 17-OHCS secretion rate increased from the basal level of 0.05 to 0.80 and 1.03 µg/kg/min.

Suzuki et al. [1975b] evaluated adrenal 17-OHCS secretion in intact and hypophysectomized dogs (anesthetized with pentobarbital) in response to cyanide hypoxia (i.v. infusion of 1.0 mg/kg potassium cyanide over 10 min). The experimental results are shown in figure 9. In intact dogs the adrenal 17-OHCS secretion rates increased from the resting levels of 0.07 ± 0.029 and 0.11 ± 0.022 µg/kg/min to the maxi-

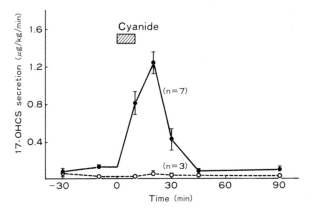

Fig. 9. Adrenocortical secretory response in 7 intact (●) and 3 hypophysectomized (○) pentobarbital-anesthetized dogs to cyanide hypoxia (i.v. infusion of 1.0 mg/kg potassium cyanide over 10 min). Adrenal 17-OHCS secretion rates are presented as mean ± SEM. [Based on data from *Suzuki* et al., 1975b.]

mum values of 1.23 ± 0.12 µg/kg/min which were observed at 20 min after the onset of the i.v. infusion of cyanide. Then the secretion rates decreased and returned to the pre-infusion levels by 45 min. The adrenal cortical secretory response to cyanide hypoxia in the dog was found to be completely abolished by hypophysectomy. The above findings were confirmed later by *Hirose* et al. [1976].

H. Hyperoxia

Adrenal ascorbic acid depletion in rats in response to hyperoxia has been reported [*Aschan*, 1953; *Gerschman and Fenn*, 1954].

Marotta et al. [1965] observed a transient insignificant decrease followed by a significant (69–95%) increase in adrenal 17-OHCS secretion in pentobarbital-anesthetized dogs which were made hyperoxic by breathing 100% O_2 at atmospheric pressure through a tracheal cannula. A significant (about 3-fold) increase in adrenal 17-OHCS secretion in pentobarbital-anesthetized dogs was also observed in their experiments during positive pressure air breathing.

In contrast, *Hale* et al. [1964] reported that peripheral plasma concentration of cortisol in man was slightly but significantly depressed during hyperoxia induced by breathing pure O_2 at atmospheric pressure.

I. Hypercapnia (Respiratory Acidosis)

Adrenal ascorbic acid and cholesterol depletions in rats following exposure to various concentrations of CO_2 in air have been demonstrated [*Langley and Kilgore*, 1955].

Richards and Pruitt [1957] exposed pentobarbital-anesthetized dogs to 20% CO_2 for 1 h in a large sealed chamber kept at 21% O_2 and atmospheric pressure. During the period of CO_2 exposure severe hypercapnia and respiratory acidosis occurred; mean arterial plasma pCO_2 was elevated to 130 mm Hg and mean arterial plasma pH was reduced to 6.94. In these dogs a marked (3- to 16-fold) increase in adrenal 17-OHCS secretion was observed.

Similar experiments had been performed in conscious [*Richards and Stein*, 1956] and pentobarbital-anesthetized [*Richards and Stein*, 1956, 1957] dogs. Conscious and anesthetized dogs were exposed in a recompression chamber at atmospheric pressure to 2.5, 5.0 and 10% CO_2 successively each for 1 h. Anesthetized dogs were also exposed to 10, 20 and 30% CO_2 each for 1 h and to 20% CO_2 for 4 h. Adrenal 17-OHCS secretion was increased in some dogs during exposure to 2.5–10% CO_2 and in all dogs exposed to 20% CO_2. An increase in adrenal 17-OHCS secretion was observed in 4 out of 10 anesthetized dogs (40%) with arterial plasma pCO_2 of 40–70 mm Hg and arterial plasma pH of 7.20–7.40, in 8 of 13 anesthetized dogs (62%) with pCO_2 of 70–100 mm Hg and pH of 7.00–7.20, and in all anesthetized dogs with pCO_2 of 100 mm Hg or higher and pH of 7.00 or less. Thus, adrenocortical stimulation by CO_2 exposure was found to be correlated with elevated pCO_2 and decreased pH. Adrenal 17-OHCS secretory response in anesthetized dogs to CO_2 exposure (20% CO_2 for 2 h) was abolished by hypophysectomy. In contrast, *Lau* [1971a] found a significant depression of adrenal 17-OHCS secretion in pentobarbital-anesthetized dogs breathing a gas mixture of 10% CO_2, 20% O_2 and 70% N_2.

In rats, breathing hypercapnic gas mixture (10% CO_2 + 20% O_2 in N_2) for 1 h significantly elevates plasma corticosterone levels [*Marotta* et al., 1976].

Metabolic Acidosis and Metabolic Alkalosis

Richards [1957] has evaluated the effect of metabolic acidosis on adrenocortical secretion in pentobarbital-anesthetized dogs. After i.v. infusion of 100–200 ml of 0.3 M HCl over 5 min, arterial plasma pH was reduced to levels less than 7.00 in 6 of 7 dogs and adrenal 17-OHCS secretion was markedly (4- to 30-fold) increased. It was also reported by him that adrenocortical secretion in anesthetized dogs was stimulated by metabolic alkalosis when arterial plasma pH was elevated to 7.60 or above and plasma bicarbonate concentration was increased.

J. Anaphylactic Shock

Suzuki et al. [1966a] evaluated adrenal 17-OHCS secretion in conscious and pentobarbital-anesthetized dogs in response to anaphylactic shock. The animals were sensitized by pretreatment with horse serum, i.e. they were injected s.c. with 0.5 ml/kg of horse serum and were injected i.v. on the following day with the same dose of horse serum. Observations were performed about 4 weeks thereafter. To induce anaphylactic shock, the sensitized animals were injected i.v. with horse serum diluted 1:1 with 0.9% saline solution at a rate of 2 ml/min until the femoral artery pulse became feeble. In 4 conscious dogs adrenal 17-OHCS secretion was found to be markedly increased. The resting 17-OHCS secretion rates were 0.01–0.21 µg/kg/min and the maximal secretion rates after inducing anaphylactic shock were 1.00–1.36 µg/kg/min, which were attained at 10–20 min after the start of horse serum injection in 2 of 4 dogs and at the end of the 2-hour period of observation in the other 2 dogs (table III). A similar finding was obtained in pentobarbital-anesthetized dogs. Adrenal 17-OHCS secretion rates were increased by anaphylactic shock from the basal values of 0–0.24 to 0.84–1.61 µg/kg/min (table III). The experimental results indicate that pentobarbital anesthesia does not block the adrenocortical secretory response in the dog to anaphylactic shock.

Aikawa et al. [1981a] have recently demonstrated that a significant elevation of adrenal secretion of cortisol and corticosterone in pentobarbital-anesthetized dogs, which is markedly reduced but not completely abolished by hypophysectomy plus bilateral nephrectomy, is induced by anaphylactic shock.

Table III. Effect of anaphylactic shock on adrenocortical secretion in conscious and pentobarbital-anesthetized dogs[1]

Dog No.	Adrenal 17-hydroxycorticosteroid secretion rate, µg/kg/min							
	minutes before		minutes after inducing anaphylactic shock					
	30	10	10	20	40	60	90	120
Conscious dogs								
1	0.01	0.03	0.66	1.00	0.14	0.08	0.04	0.12
2	0.03	0.02	0.47	0.96	1.17	1.04	1.22	1.36
3	0.09	0.21	0.15	0.43	0.53	0.78	0.66	1.06
4	0.09	0.16	1.29	0.77	0.73	0.76	0.23	0.27
Pentobarbital-anesthetized dogs								
5	0.11	0.23	1.26	1.61	1.11	0.20	0.02	0.06
6	0.24	0.23	1.37	1.34	1.55	1.38	1.35	0.81
7	0.00	0.04	0.56	0.84	0.41	0.06	0.08	0.20

[1] Reproduced, with permission, from Suzuki et al. [1966a].

K. Histamine

Systemic administration of histamine in rats induces a significant depletion of adrenal ascorbic acid [Sayers and Sayers, 1947; Gray and Munson, 1951; Nasmyth, 1951, 1955; Tepperman et al., 1951; Briggs and Munson, 1955; Arimura, 1955/56; Olling and De Wied, 1956; Van Peenen and Way, 1957; Casentini et al., 1959; Rochefort et al., 1959; Mitamura, 1960b; Giuliani et al., 1961; Leeman et al., 1962; Wexler, 1963; Barrett and Stockham, 1965].

Endröczi et al. [1959] have attempted to evaluate adrenocortical secretion in response to histamine in anesthetized cats by a direct method. They observed a significant increase in adrenal corticosteroid secretion after s.c. injection of histamine (3.0 mg/kg). However, the time course of adrenocortical secretion following histamine injection was not elucidated in their study.

Suzuki et al. [1963] directly evaluated the adrenal 17-OHCS secretion in conscious dogs in response to histamine. In their study a marked increase in adrenal 17-OHCS secretion was observed following i.v. injection of histamine in doses of 0.05–1.0 mg/kg; the maxi-

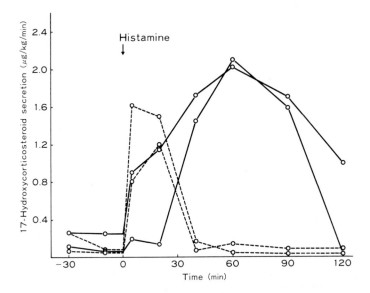

Fig. 10. Time course of adrenal 17-OHCS secretion in conscious dogs i.v. injected with histamine dihydrochloride in doses of 0.2 (O- - -O) and 1.0 (O———O) mg/kg. [Based on data from *Suzuki* et al., 1963.]

mum rates of secretion of 17-OHCS (0.76–1.62 μg/kg/min) were attained at 5 or 20 min in dogs injected with 0.05–0.2 mg/kg histamine and they (2.02 and 2.10 μg/kg/min) were attained at 60 min when injected with 1.0 mg/kg histamine. Duration of increased adrenal 17-OHCS secretion following administration of histamine was 20, 40 and 90 min or longer in dogs injected with 0.05–0.2, 0.5 and 1.0 mg/kg of histamine, respectively. The time courses of adrenal 17-OHCS secretion following i.v. injection of 0.2 and 1.0 mg/kg histamine are presented in figure 10. Subsequently, *Papp* et al. [1964] and *L'Age* et al. [1970] also showed an increase in adrenal secretion of cortisol in conscious dogs following i.v. injection (or infusion) of histamine.

    Asano [1966] measured adrenal 17-OHCS secretion in pentobarbital-anesthetized dogs in response to histamine. The adrenocortical secretory response to 0.05 mg/kg histamine (i.v.) was found to be completely abolished by hypophysectomy, thus suggesting that the effect of histamine on adrenocortical secretion is mediated through the pituitary. *Tanigawa* [1967] studied the mechanism of stimulatory action of

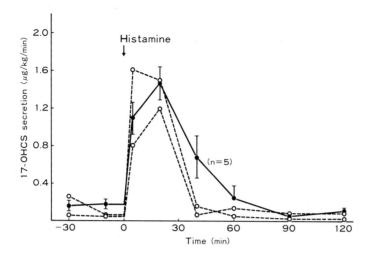

Fig. 11. Adrenal 17-OHCS secretion rates (mean ± SEM) in 5 pentobarbital-anesthetized dogs (●) before and after i.v. injection of 0.2 mg/kg histamine dihydrochloride. For comparisons the data in 2 conscious dogs (○) i.v. injected with the same dose of histamine are also presented in the figure. [Based on data from *Tanigawa*, 1967, and *Suzuki* et al., 1963.]

histamine on adrenocortical secretion. In pentobarbital-anesthetized dogs a marked increase in adrenal 17-OHCS secretion was observed after i.v. injection of histamine. As shown in figure 11, the magnitude of adrenocortical secretory response to 0.2 mg/kg histamine (i.v.) in pentobarbital-anesthetized dogs was found to be comparable with that in conscious dogs reported by *Suzuki* et al. [1963]. Adrenocortical secretory response to histamine was not affected by bilateral section of the splanchnic nerves; it was found to be markedly depressed but not always completely abolished by hypophysectomy.

Katsuki et al. [1967], *Narita* [1971] and *Yamashita* et al. [1973] also observed in pentobarbital-anesthetized dogs a significant increase in adrenal 17-OHCS secretion following i.v. injection of histamine. *Narita* [1971] evaluated the adrenal 17-OHCS secretion in response to small doses of histamine, such as 0.001–0.005 mg/kg. An i.v. injection of histamine in a dose of 0.005 or 0.002 mg/kg was found to be effective, whereas histamine in a dose of 0.001 mg/kg was ineffective, in elevating significantly the adrenal 17-OHCS secretion.

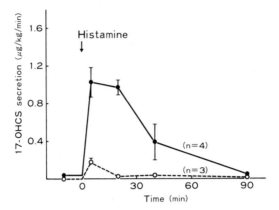

Fig. 12. Adrenal 17-OHCS secretion (mean ± SEM) in 4 intact (●) and 3 hypophys-ectomized (○) pentobarbital-anesthetized dogs in response to histamine (i.v. injection of 0.1 mg/kg histamine dihydrochloride). [Based on data from *Hirose* et al., 1976.]

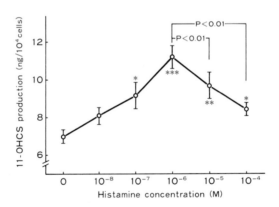

Fig. 13. Production in vitro of 11-OHCS (mean ± SEM) by trypsin-dispersed adrenal cells of guinea pigs in response to histamine. * $p < 0.02$, ** $p < 0.01$, *** $p < 0.001$ vs control (Student's t-test). [Reproduced, with permission, from *Matsumoto* et al., 1981.]

The finding that adrenal 17-OHCS secretion in the dog in response to histamine (0.1 mg/kg, i.v.) is not always completely abolished, although markedly reduced, by hypophysectomy [*Tanigawa*, 1967] has subsequently been confirmed [*Hirose* et al., 1976] (fig. 12). A small but significant increase in adrenal cortisol and corticosterone secretion in

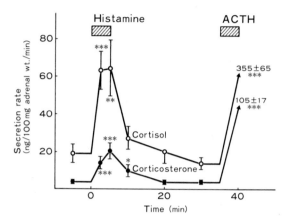

Fig. 14. Secretion rates (mean ± SEM) of cortisol and corticosterone of perfused adrenal glands of hypophysectomized-nephrectomized dogs (n = 7) before and after 5-min infusions of 0.5 mg histamine dihydrochloride and 0.1 U ACTH. * p < 0.05, ** p < 0.02, *** p < 0.01 vs the secretion rate before infusion of histamine or ACTH (Student's t-test). [Based on data from *Aikawa* et al., 1981b.]

hypophysectomized dogs [*Hirose* et al., 1977] and in hypophysectomized-nephrectomized dogs [*Aikawa* et al., 1979] after i.v. administration of histamine (0.1 mg/kg) has been presented. Thus, a possibility is suggested that the stimulatory effect of histamine on adrenocortical secretion might in a small part be due to a direct action of histamine on the adrenal cortex. The direct stimulatory effect of histamine on the adrenal cortex has recently been demonstrated in the in vitro experiments with dispersed dog [*Hirose* et al., 1978, 1979] and guinea pig [*Matsumoto* et al., 1981] adrenal cells (fig. 13), and in experiments with perfused dog adrenal gland [*Aikawa* et al., 1981b] (fig. 14).

An elevation in plasma corticosterone levels in rats [*Yates* et al., 1961; *Leeman* et al., 1962; *Smelik*, 1963; *Barrett and Stockham*, 1965; *Arimura* et al., 1967; *Stark* et al., 1968; *Hirai* et al., 1970; *Butte* et al., 1976] and in mice [*Kakihana* et al., 1974] following systemic administration of histamine has been reported. An increase in corticosteroid production in vitro of excised adrenal glands in rats injected systemically with histamine [*Smelik*, 1963; *De Wied*, 1964] and elevations of adrenal corticosterone levels in rats in response to histamine [*Holzbauer*, 1957; *Barrett and Stockham*, 1965; *Hirai* et al., 1970] have also been reported.

L. Hemorrhage

Sayers et al. [1945] have observed hemorrhage-induced depletions of adrenal ascorbic acid and cholesterol in rats, which are completely abolished by hypophysectomy. Adrenal ascorbic acid depletion in response to hemorrhage was also reported by *Ronzoni and Reichlin* [1950].

In the study by *Frank* et al. [1955], the adrenal vein was cannulated in dogs under ether anesthesia and observations were started soon after recovery from anesthesia. Following hemorrhage (removal of blood in volumes of 20–51 ml/kg) the blood pressure was reduced to 30 mm Hg and the adrenal blood flow was depressed to 3–26% of prehemorrhage values. The adrenal corticosteroid secretion rates decreased in 5 out of 13 dogs below 25%, in 5 dogs to 56–77% of the values before hemorrhage and in 3 dogs the secretion rates remained unchanged or somewhat increased. In their study, however, experiments were conducted before the animals were recovered from the effects of surgical trauma of adrenal vein cannulation, and hence under maximal adrenal stimulation by endogenous ACTH.

In pentobarbital-ether-anesthetized dogs in which the observations were started shortly after surgery of adrenal vein cannulation, *Ganong* et al. [1955a] failed to find any definite change in adrenal 17-OHCS secretion following hemorrhage even when the blood pressure and adrenal blood flow were markedly reduced.

In contrast, *Hume and Nelson* [1955a] observed in some of pentobarbital-anesthetized dogs, which were subjected to surgical trauma of adrenal vein cannulation, a marked increase in adrenal 17-OHCS secretion following hemorrhage. In 1 dog, whose blood pressure did not fall below 64 mm Hg after hemorrhage, the adrenal 17-OHCS secretion rate increased from the prehemorrhage level of 9.0 to 21.1 µg/min. In another dog, whose blood pressure fell to 35 mm Hg after hemorrhage, the adrenal 17-OHCS secretion rate decreased from the prehemorrhage value of 5.0 to 2.9 µg/min. In the latter dog, however, the blood pressure rose over the next 2 min to 58 mm Hg and the adrenal 17-OHCS secretion rate increased to 23.8 µg/min.

Walker et al. [1959a] noted that the adrenal 17-OHCS secretion rates increased by 13.3–81.3% in 3 out of 4 dogs and decreased by 20.6% in 1 dog following removal of blood in volumes of 10–38 ml/kg. In their study, experiments were started soon after recovery from anesthesia and the 17-OHCS secretion rates before hemorrhage were 3.4–8.3 µg/min.

Mulrow and Ganong [1961] evaluated adrenal secretions of 17-OHCS and corticosterone in response to hemorrhage (removal of 15 ml/kg of blood within 6 min) in 5 pentobarbital-anesthetized dogs whose adrenal veins were cannulated on the day prior to the experiment. In 3 dogs with low prehemorrhage secretion rates of 17-OHCS and corticosterone, such as 0.3–1.5 and 0.13–0.18 µg/min, respectively, marked elevations of both 17-OHCS and corticosterone secretions were observed following hemorrhage. Maximal secretion rates in these dogs after hemorrhage were 4.3–10.8 µg/min for 17-OHCS and 0.90–1.8 µg/min for corticosterone, respectively. In contrast, in 2 dogs with high prehemorrhage adrenal secretion rates of 17-OHCS and corticosterone, such as 6.5 and 12. 6 µg/min and 1.3 and 1.4 µg/min, respectively, hemorrhage induced little change in the secretion rates. In 8 out of 9 hypophysectomized dogs no definite changes in 17-OHCS secretion rates were observed following hemorrhage. In 1 dog, however, a slight increase in 17-OHCS secretion was observed. Adrenal secretion of corticosterone was evaluated in 4 hypophysectomized dogs and was found to be slightly increased after hemorrhage.

Gann and Egdahl [1965] reported that the adrenal 17-OHCS secretion in pentobarbital-anesthetized dogs was increased from the basal secretion rates of 1.5 ± 0.6 to 20.6 ± 3.2 µg/min following hemorrhage sufficient to reduce mean arterial pressure to 60 mm Hg; following rapid reinfusion of shed blood into dogs subjected to hemorrhage, the secretion rates returned to the basal levels within 15 or 30 min.

In pentobarbital-anesthetized dogs the adrenocortical secretory response to hemorrhage is proportional to the logarithm of the volume of shed blood [*Gann*, 1966].

Johnson et al. [1971] examined the effect of continuous progressive hemorrhage on adrenocortical secretion in pentobarbital-anesthetized dogs. In dogs whose adrenal vein was cannulated shortly before the experiment, adrenal secretion of cortisol reduced slowly and progressively during the continuous hemorrhage period. In dogs in which adrenal vein cannulation was done 3 days before the experiment, the adrenal cortisol secretion rate increased at first progressively, attained maximal values and then progressively decreased toward the basal levels during the continuous hemorrhage period.

Gann et al. [1977] evaluated adrenal cortisol secretion in pentobarbital-anesthetized dogs in response to each of two sequential hemorrhages with a time interval of 90 min. In experiments in which the effect of each of two sequential hemorrhages of 10 ml/kg per 1.5 min was evaluated, the adrenal cortisol secretory response to the second hemorrhage was significantly smaller than that to the first hemorrhage. In experiments in which the animals were expanded by i.v. infusion of 6% dextran in saline and the effect of each of two sequential hemorrhages of 10 ml/kg per 45 s was evaluated, the secretory response to the second hemorrhage was found to be significantly greater than that to the first hemorrhage. These findings suggest that physiological inhibition and facilitation on the pituitary-adrenocortical activity might be produced by hemorrhage.

Atkins and Marotta [1963] studied the effect of hemorrhage on peripheral plasma 17-OHCS levels in pentobarbital-anesthetized dogs. A 10% hemorrhage (rapid removal of blood in volumes of 10% of the total blood) increased plasma concentrations of 17-OHCS from the basal levels of 8.0 ± 1.1 to 15.8 ± 1.5 µg/100 ml and an additional 15% hemorrhage induced a further increase in plasma 17-OHCS to 22.3 ± 1.3 µg/100 ml. In dogs subjected to 35% hemorrhage at the rate of 5% hemorrhage every 10 min, a significant elevation of plasma 17-OHCS levels was first observed after 25% hemorrhage. A rise of plasma corticosterone levels in pentobarbital-anesthetized rats following hemorrhage has also been reported [*Feldman* et al., 1975a].

M. Surgical Stress

Adrenal ascorbic acid depletion in response to surgical stress in rats anesthetized with pentobarbital [*Briggs and Munson*, 1955; *Munson and Briggs*, 1955; *Ohler and Sevy*, 1956; *Sevy* et al., 1957; *Wexler*, 1963; *Brodish*, 1964] or ether [*Abelson and Baron*, 1952; *McCann*, 1953; *McCann and Haberland*, 1960; *Vernikos-Danellis and Marks*, 1962; *McCann* et al., 1966] has been reported. In pentobarbital-anesthetized cats surgical stress induces a small but significant depletion of adrenal ascorbic acid [*Schwartz and Kling*, 1960].

Ganong et al. [1955a] observed that adrenal 17-OHCS secretion was elevated in pentobarbital-ether-anesthetized dogs following adrenal vein cannulation to the maximal levels which was not additionally increased by i.v. injection of 10 mU ACTH; mean adrenal 17-OHCS secretion rates before and after ACTH injection were 8.6 ± 0.7 and 8.9 ± 1.0 µg/min, respectively. Similar results were obtained by *Richards and Pruitt* [1957].

In experiments by *Hume and Nelson* [1955a] the adrenal 17-OHCS secretion rates in dogs subjected to adrenal vein cannulation under ether anesthesia were 9.0–22.2 µg/min and those under pentobarbital anesthesia were 2.4–20.8 µg/min. After the second postoperative day the secretion rates were 0–2.0 µg/min. The experimental results might suggest that adrenocortical secretion in dogs is markedly elevated by surgical stress. Similar findings were reported by *Nelson* et al. [1956], *Hume* [1957], *Egdahl* [1964] and *Katsuki* et al. [1967]. *Zukoski and Ney* [1966] also noted a marked increase in adrenal 17-OHCS secretion in pentobarbital-anesthetized dogs in response to surgical stress (laparotomy); the mean secretion rate of 17-OHCS for 8 controls was 1.6 ± 0.8 µg/min and that following laparotomy was 7.2 ± 2.2 µg/min.

Hume et al. [1962] performed direct evaluation of adrenal 17-OHCS secretion in man during surgery and in postoperative period. The mean secretion rates of 17-OHCS during surgery and in the postoperative period were 33.1 and 3.1 µg/min, respectively.

A significant elevation of plasma corticosterone levels in pentobarbital-anesthetized rats in response to surgical stress has been reported [*Yates* et al., 1961; *Leeman* et al., 1962; *Dallman and Jones*, 1973]. The rise of peripheral plasma levels in man of 17-OHCS [*Franksson and Gemzell*, 1953; *Sandberg* et al., 1954; *Tyler* et al., 1954; *Virtue* et al., 1957; *Hammond* et al., 1958; *Steenburg* et al., 1961; *Estep* et al., 1963; *Oyama* et al., 1968], 11-OHCS [*Nilsson* et al., 1963; *Asfeldt and Elb*, 1968; *Plumpton and Besser*, 1969; *Kehlet* et al., 1973], cortisol [*Lewis*, 1963; *Blichert-Toft* et al., 1967; *Hamanaka* et al., 1970; *Jensen and Blichert-Toft*, 1971] and corticosterone [*Hamanaka* et al., 1970] during and following surgery has been observed.

N. Exposure to Cold

A number of investigators [*Long*, 1947; *Sayers and Sayers*, 1947; *Kimura*, 1954; *Arimura*, 1955/56; *Nowell*, 1959; *Maickel* et al., 1961; *Gaunt* et al., 1962; *Smith* et al., 1963] have observed a significant depletion of adrenal ascorbic acid in rats following exposure to cold. Adrenal cholesterol depletion in rats exposed to cold has also been reported [*Long*, 1947]. Adrenal ascorbic acid [*Long*, 1947; *Maickel* et al., 1961] and cholesterol [*Long*, 1947] depletions in rats in response to cold stress are blocked by hypophysectomy. Adrenal ascorbic acid [*Long*, 1947] and cholesterol [*Long*, 1947; *Tang and Patton*, 1951] depletions in guinea pigs after exposure to cold have also been noted.

Adrenal cortical secretion in dogs in response to cold exposure has been studied [*Nelson* et al., 1956]. Conscious dogs were placed in a cold room at −10 °C for periods of 1–33 h. Under these conditions the animal showed no definite fall in body temperature. The secretion rate of 17-OHCS was 2.8 ± 0.6 µg/min before cold exposure and it was 2.1 ± 0.4 µg/min during exposure to cold. However, the 17-OHCS secretion rate was found to be elevated during exposure to room temperature following cold exposure; it increased to 7.1 ± 1.1 µg/min. *Egdahl and Richards* [1956b] evaluated directly the effect of extreme cold exposure on adrenocortical secretion in the dog. In their experiments 6 and 4 conscious dogs were placed in a cold room of −46 to −50 °C for 2–5 or 25–28 h and of −75 to −79 °C for 4–5 h, respectively. During exposure to cold, hypothermia was not observed. Adrenal 17-OHCS secretion was markedly increased soon after the onset of cold exposure and remained elevated for 1–3 h; after that it returned to the basal levels. Mean secretion rates of 17-OHCS before cold exposure were 0.7–4.2 µg/min and maximum secretion rates during cold exposure were 12.2–29.2 µg/min.

Maickel et al. [1961] observed that plasma corticosterone levels in rats were significantly elevated after 1–8 h of cold exposure (4 °C); the maximal elevation (about 300% increase) was found after a 2-hour cold exposure. Subsequently, *Smith* et al. [1963], *Jobin* et al. [1976] and *Takeuchi* et al. [1977] also reported a marked elevation of plasma corticosterone levels in rats in response to cold stress.

Chowers et al. [1964] showed a significant elevation in plasma cortisol levels in dogs whose environmental temperature was suddenly lowered to 0 °C. When the environmental temperature was reduced gradually over 45-min period to 0 °C, plasma cortisol levels did not show any significant changes. An elevation in plasma 17-OHCS levels in cold-exposed dogs has been reported also by *Sadowski* et al. [1972, 1975].

O. Exposure to Heat

Significant depletions of adrenal ascorbic acid [*Sayers and Sayers*, 1947; *Langley and Kilgore*, 1955; *Degonskii*, 1977] and cholesterol [*Langley and Kilgore*, 1955] in rats following exposure to heat have been reported. *Richards and Egdahl* [1956] have determined the effect of hyperthermia induced by exposure to heat on adrenal 17-OHCS secretion in pentobarbital-anesthetized dogs. After the animals were immersed to the shoulders in a water bath (36–38 °C), the temperature of the bath water was elevated within 2–3 min to 50 °C and maintained at this level for 40–50 min. The rectal temperature was raised to 42 °C within the first 15–20 min of the immersion period and a marked increase in the adrenal 17-OHCS secretion rate was observed. When the rectal temperature was elevated further and attained 44–45 °C, adrenal blood flow markedly decreased as a result of circulatory failure and adrenal 17-OHCS secretion returned to the baseline levels. In another series of experiments, when the temperature of the bath water was slowly elevated to attain 40–41 °C and the rectal temperature of the animals was raised to 39–40 °C, a marked increase in adrenal 17-OHCS secretion was observed in most cases. However, when the rectal temperature was further elevated by elevating the bath water temperature, adrenal 17-OHCS secretion was found to be rather decreased. Adrenal 17-OHCS secretion in anesthetized dogs in response to hyperthermia was completely abolished by hypophysectomy, thus suggesting that it is mediated by the pituitary gland.

Chowers et al. [1966] studied the effect of exposure to heat in dogs on peripheral plasma cortisol levels. A small but significant elevation in plasma cortisol levels was observed when environmental temperature was suddenly elevated to 42–45 °C, whereas only a slight yet significant elevation of plasma cortisol levels was observed when environmental temperature was raised gradually from 25 to 45 °C during a 45-min period.

P. X-Irradiation

Adrenal ascorbic acid [*North and Nims*, 1949; *Venters and Painter*, 1951; *Wexler* et al., 1952; *Hochman and Bloch-Frankenthal*, 1953; *Oster* et al., 1953; *Bacq and Fischer*, 1957; *Bacq* et al., 1960; *Binhammer and*

Crocker, 1963] and cholesterol [*Patt* et al., 1947, 1948; *North and Nims,* 1949; *Nims and Sutton,* 1954; *Bacq and Fischer,* 1957; *Bacq* et al., 1960] depletions have been observed in rats after total body irradiation with X-ray. A decrease in adrenal cholesterol content in rats following whole body X-ray irradiation is blocked by hypophysectomy [*Patt* et al., 1948; *Nims and Sutton,* 1954]. *Wexler* et al. [1955] reported that adrenal ascorbic acid response in rats to half-body X-irradiation was much greater than that to total body X-irradiation.

Van Cauwenberge et al. [1957a, b] have shown a significant elevation of plasma corticosteroid or 17-OHCS levels in rats in response to total body X-irradiation. *Eechaute* et al. [1962a] found that plasma corticosterone levels and corticosterone production in vitro of excised incubated adrenal glands in rats were elevated at 2.5 and 72 h after whole body X-irradiation. *Hameed and Haley* [1964] demonstrated that plasma and adrenal corticosterone levels in intact rats were significantly elevated at 2.5 and 48 h after total body X-irradiation, at 72 h after localized head X-irradiation and at 2.5 h after localized body X-irradiation. In hypophysectomized rats, however, they failed to find any elevation in either plasma or adrenal corticosterone levels following total body or localized head or localized body X-irradiation.

French et al. [1955] have attempted to determine the effect of various doses of total body X-irradiation in rhesus monkeys on peripheral plasma 17-OHCS levels. No plasma 17-OHCS response was observed following 50 R X-irradiation. After 400–800 R X-irradiation plasma 17-OHCS levels were markedly elevated, attained a peak at 4–8 h, returned to the preirradiation levels within 12 h, and were again elevated shortly before death (6–16 days after irradiation). Similar findings in rhesus monkeys were reported by *Wolf and Bowman* [1964].

In intact but not in hypophysectomized pentobarbital-anesthetized dogs adrenal 11-OHCS secretion has been found to be elevated by localized adrenal irradiation with 800–2,000 R of X-ray [*Nakasone,* 1972]. In man *Notter and Gemzell* [1956] observed a fall followed by an elevation of peripheral plasma 17-OHCS levels after localized adrenal X-irradiation (50–400 R).

Shimizu et al. [1973] have shown in pentobarbital-anesthetized dogs a marked increase in adrenal 17-OHCS secretion in response to 500 R localized head X-irradiation and a marked increase preceded by a temporary decrease following 1,000 R localized head X-irradiation. Adrenocortical secretory responsiveness to histamine [*Yamashita* et al., 1973] and acetylcholine [*Yamashita* et al., 1978] has been found to be significantly depressed in pentobarbital-anesthetized dogs by localized head irradiation with 1,000 R of X-ray on the day before observation.

Q. Ingestion of Food

The effect of food ingestion on adrenocortical secretion in dogs has been evaluated by *Suzuki* et al. [1971]. In their experiments adrenal 17-OHCS secretion in conscious dogs was only slightly and transiently increased following ingestion of food (250–300 g of boiled beef with a small amount of milk). The secretion rate before food ingestion was 0.04–0.25 µg/kg/min and the maximum secretion rate after food ingestion was 0.36–0.55 µg/kg/min.

In rats *Heybach and Vernikos-Danellis* [1979] observed that ingestion of food and water resulted in a depression of plasma corticosterone levels which had been elevated by a 23-hour/day deprivation of food and water for 21 days. In rats fed and watered ad libitum, however, no significant change in plasma corticosterone levels was observed following ingestion of food and water.

Chapter 3. Adrenal Cortical Secretion in Response to a Variety of Chemical Substances Involving Hormones, Neuropharmacologic Agents and Bacterial Products

A. Insulin

An increase in pituitary-adrenocortical secretory activity in rats following administration of insulin, as judged by adrenal ascorbic acid depletion, is well documented [*Gershberg and Long*, 1948; *Dury*, 1950; *Ronzoni and Reichlin*, 1950; *Hetzel and Hine*, 1951; *Vogt*, 1951b; *Mitamura*, 1960b; *Giuliani* et al., 1961].

Goldfien et al. [1958] were the first to evaluate directly the adrenocortical secretory response to insulin hypoglycemia. They observed a marked increase in adrenal 17-OHCS secretion in 3 out of 5 pentobarbital-anesthetized dogs following i.v. injection of insulin (1.0 U/kg). *Suzuki* et al. [1964b] studied the effect of insulin hypoglycemia on adrenocortical secretion in 5 conscious dogs. The adrenal 17-OHCS secretion rate was found to be elevated without exception following administration of insulin (10 U/kg, i.v.). It increased from the basal levels of 0.04–0.14 to 0.55–3.65 µg/kg/min. *Suzuki* et al. [1965c] noted that the adrenal 17-OHCS secretion rate in conscious dogs, whose spinal cord had been transected at the level of C_{7-8} 2 days previously under pentobarbital anesthesia, was increased after i.v. injection of insulin (10 U/kg) from the basal levels of 0.02–0.10 to 0.95–2.78 µg/kg/min.

Zukoski [1964] showed that the adrenal 17-OHCS secretion rate in conscious dogs was raised following i.v. injection of insulin (0.5 U/kg) and it returned to the basal levels by 60–100 min even though there still existed a severe hypoglycemia. Experimental results in pentobarbital-anesthetized dogs did not differ from those in conscious dogs. The same author [*Zukoski*, 1966] reported that a marked elevation of adrenal 17-OHCS secretion rate in pentobarbital-anesthetized dogs, which returned to the basal levels by 120 min, was induced by 0.5 but not 0.1 U/kg of insulin. Bilateral supradiaphragmatic vagotomy failed to impair the adrenal 17-OHCS secretion in response to 0.5–1.0 U/kg insulin. A marked increase in adrenal 17-OHCS secretion following

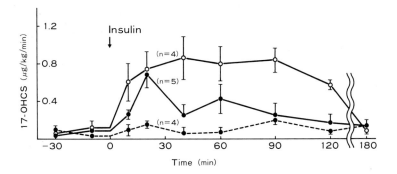

Fig. 15. Adrenal 17-OHCS secretion rates (mean ± SEM) in pentobarbital-anesthe-tized dogs i.v. injected with 1.0 (O), 0.1 (●——●) and 0.05 (●- - -●) U/kg insulin. [Based on data from *Shibata*, 1971.]

i.v. injection of insulin (0.5 U/kg) in 2 conscious dogs was also noted. *Asano* [1966] also observed a marked increase in adrenal 17-OHCS se-cretion in intact, but not in hypophysectomized, pentobarbital-anes-thetized dogs following i.v. administration of insulin.

Matsui and Plager [1966] made a detailed study on the mechanism of adrenocortical secretory response to insulin hypoglycemia. They collected 5-min samples of adrenal venous blood in pentobarbital-anesthetized dogs serially for the first 30 min following insulin admin-istration and alternately for the next 30 min. Experimental results showed that the point at which time an increase in adrenal 17-OHCS secretion was first observed after insulin administration was correla-tive with the rate of fall in blood glucose levels rather than absolute blood glucose concentrations. It was also shown that the adrenocorti-cal response to insulin hypoglycemia was completely abolished by hy-pophysectomy. *Shibata* [1971] evaluated adrenal 17-OHCS secretion in pentobarbital-anesthetized dogs in response to various i.v. doses of in-sulin. Insulin in doses of 0.1–1.0 U/kg was found to be effective in stimulating adrenocortical secretion, whereas 0.05 U/kg insulin was ineffective (fig. 15). *Fujii* [1972] noted an increase in adrenal 11-OHCS secretion in pentobarbital-anesthetized dogs following i.v. administra-tion of insulin (1.0 U/kg).

A significant elevation of plasma corticosterone levels in rats is in-duced by insulin hypoglycemia [*Kraicer and Logothetopoulos*, 1963b].

An elevation of peripheral plasma 17-OHCS levels [*Bliss* et al., 1954; *Froesch* et al., 1954; *Froesch*, 1955; *Christy* et al., 1957; *Weinges and Schwarz*, 1960; *Werk* et al., 1961] and in peripheral plasma cortisol (or 11-OHCS) levels [*Amatruda* et al., 1960; *Arner* et al., 1962; *Landon* et al., 1963, 1965; *Plumpton and Besser*, 1969; *Müller-Hess* et al., 1974; *Plonk* et al., 1974; *Hökfelt* et al., 1978] in man following administration of insulin has been observed.

Alloxan

It is well established that an initial hyperglycemia which is followed by a prolonged hypoglycemic phase is observed after systemic administration of alloxan. Alloxan-induced hypoglycemia seems to be attributed to insulin release from degenerating β-cells of the islets of Langerhans in the pancreas.

A significant depletion of adrenal ascorbic acid in rats has been observed at 2 h [*Kosaka*, 1954] and during the observation period from 2 to 24 h [*Van Cauwenberge*, 1956] after administration of alloxan. Adrenal cholesterol depletion in rats in response to alloxan has also been noted [*Van Cauwenberge*, 1956].

Suzuki et al. [1966b] studied the effect of alloxan on adrenal 17-OHCS secretion in conscious dogs. The adrenal 17-OHCS secretion rates, which were 0.09–0.31 μg/kg/min before injection, increased 20–40 min after injection of alloxan (100 mg/kg, i.v.) to 0.50–1.26 μg/min in 4 out of 5 dogs, then decreased and again increased at 1–3 h to the levels of 0.88–2.35 μg/kg/min in 4 out of 5 dogs.

Saba and Hoet [1962] observed that plasma corticosterone levels in rats were significantly elevated at 30 and 60 min after i.v. injection of alloxan (40 mg/kg). *Kraus* [1973] noted significant elevations of plasma and adrenal corticosterone levels in rats 72 h after i.p. injection of alloxan (20 mg/100 g).

B. Vasopressin

A number of investigators have observed that adrenal ascorbic acid in rats is significantly reduced by a systemic administration of Pituitrin (a commercial preparation of posterior pituitary extract) [*Kimura*, 1954], Pitressin (a commercial preparation of pressor principle of posterior pituitary extract) [*Nagareda and Gaunt*, 1951; *Briggs and*

Munson, 1955; *Arimura*, 1955/56; *Itoh*, 1957; *De Wied* et al., 1958, 1961; *Casentini* et al., 1959; *Nowell*, 1959; *Giuliani* et al., 1961; *Leeman* et al., 1962] and vasopressin [*Itoh*, 1957; *Sevy* et al., 1957; *Casentini* et al., 1959; *Chauvet and Acher*, 1959; *De Wied* et al., 1961; *Leeman* et al., 1962]. Adrenal ascorbic acid depletion in rats following administration of Pitressin [*Arimura*, 1955/56; *De Wied* et al., 1958; *Casentini* et al., 1959] and vasopressin [*Casentini* et al., 1959] is abolished by hypophysectomy.

Katsuki [1961] showed an increase in adrenal 17-OHCS secretion in the dog following i.v. infusion of vasopressin (5–20 U over 10 min). *Asano* [1966] also noted a marked elevation of adrenal 17-OHCS secretion rate in pentobarbital-anesthetized dogs induced by i.v. injection of vasopressin (0.25–0.5 U/kg). In his experiments the adrenocortical secretory response to vasopressin was found to be markedly reduced but not completely abolished by hypophysectomy.

Leeman et al. [1962] observed a significant elevation of plasma corticosterone levels in rats following injection of Pitressin, a response completely abolished by hypophysectomy. Peripheral plasma levels of 17-OHCS in conscious cats [*Krieger* et al., 1968; *Krieger and Rizzo*, 1969] and peripheral plasma 11-OHCS levels in conscious dogs [*Malayan* et al., 1980] have been found to be elevated by i.v. administration of vasopressin. In conscious trained dogs *Nichols and Guillemin* [1959] observed a rise in peripheral plasma 17-OHCS levels following injection of vasopressin into the carotid artery. However, the threshold dose of vasopressin for plasma 17-OHCS response was found to be 3.5 × 10³ to 7.1 × 10³ times the dose necessary for maximal antidiuresis.

Peripheral plasma corticosteroid response in man to vasopressin has been studied by a number of investigators. The rise of plasma 17-OHCS [*McDonald and Weise*, 1956; *Clayton* et al., 1963; *Librik and Clayton*, 1963], 11-OHCS [*Gwinup*, 1965; *Brostoff* et al., 1968; *Hortling* et al., 1971; *Toft* et al., 1971], cortisol [*Van der Wal* et al., 1961; *Takebe* et al., 1968; *Tucci* et al., 1968; *Andersson* et al., 1972] and corticosterone [*Van der Wal* et al., 1961] following administration of vasopressin (or Pitressin) has been reported.

A number of studies have been done to elucidate the site or mode of action of vasopressin in stimulating adrenocortical secretion. It has been suggested that adrenocortical secretory response to vasopressin is mediated by pituitary ACTH release, but evidence for direct stimulation by vasopressin on the adrenal cortex has also been provided.

Royce and Sayers [1958b] found a significant depletion of adrenal ascorbic acid in hypophysectomized rats following vasopressin (or Pitressin) injection. They suggested 2 alternative explanations; (1) vasopressin may directly act on the adrenal cortex or (2) adrenocortical secretory response to vasopressin may be mediated by ACTH from some extra-pituitary storage sites. *Hilton* et al. [1960] performed direct arterial perfusion of the dog adrenal gland in situ and demonstrated a direct stimulatory effect of vasopressin on the adrenal cortex; vasopressin infused into the arterial circuit leading to the adrenal gland was found to produce a marked increase in adrenal cortisol secretion. *Andersen and Egdahl* [1964] showed that adrenal 17-OHCS secretion in pentobarbital-anesthetized dogs was increased without exception by injection of vasopressin into the adrenal artery in doses of 25–100 mU. However, they also showed that the adrenocortical secretion was not stimulated by 25 mU vasopressin in all 11 dogs, by 50 mU in 6 of 11 dogs and by 100 mU in 2 of 11 dogs, when vasopressin was injected into the carotid artery. They suggested that in the dog vasopressin at low dosage levels stimulates directly, but not indirectly via pituitary ACTH release, the adrenocortical secretion. However, they also suggested a possibility that endogenous vasopressin could reach the anterior pituitary by direct pathway in concentration enough to stimulate ACTH release.

Kwaan and Bartelstone [1959] observed a significant elevation in adrenal 17-OHCS secretion rate in trained conscious dogs following injection of 2 mU vasopressin into the third ventricle, whereas they failed to find any significant change in adrenocortical secretion following i.v. injection of 20 mU vasopressin.

Evidence for a direct stimulatory action of vasopressin on pituitary ACTH release has been provided by following findings: (a) vasopressin-induced ACTH release in vitro from the rat anterior pituitary tissue [*Fleischer and Vale*, 1968; *Chan* et al., 1969; *Pearlmutter* et al., 1974b; *Buckingham and Hodges*, 1977; *Luts-Bucher* et al., 1977] and from the isolated rat pituitary cells [*Portanova and Sayers*, 1973a; *Fehm* et al., 1975; *Gillies* et al., 1978; *Gillies and Lowry*, 1980], (b) an increase in corticosterone levels of adrenal venous blood in hypophysectomized rats with a pituitary tumor following injection of minute amounts of vasopressin into the arterial blood supply of the tumor [*Grindeland and Anderson*, 1964], and (c) a marked increase in adrenal cortisol secretion in conscious dogs following infusion of vasopressin into the anterior pituitary in doses which are only minimally effective when injected i.v. [*Gonzalez-Luque* et al., 1970].

In rats whose forebrain including the median eminence has been removed, plasma corticosterone levels are significantly elevated after i.v. injection of vasopressin, suggesting that the median eminence of the hypothalamus and other forebrain structures are not essential to the adrenocortical secretory response to vasopressin and that vasopressin stimulates, at least in part, directly the pituitary [*Dunn and Critchlow*, 1971].

Hiroshige et al. [1968] and *Hiroshige* [1971] demonstrated a significant elevation of plasma corticosterone levels in the rat following microinjection of small amounts of vasopressin into the posterior but not anterior pituitary. They suggested the possibility that vasopressin primarily acts on the posterior pituitary to mobilize a substance with ACTH-releasing activity, which in turn stimulates the anterior pituitary.

Yasuda et al. [1978] observed a marked elevation of plasma corticosterone levels in response to vasopressin (or Pitressin) in hypophysectomized rats bearing three whole pituitaries (or anterior pituitaries) transplanted under the kidney capsule but not in hypophysectomized rats without pituitary transplantation. This plasma corticosterone response to vasopressin in rats with transplanted pituitaries (or anterior pituitaries) was not affected by hypothalamic destruction, suggesting that vasopressin may act in vivo directly on the anterior pituitary to stimulate ACTH release and the basal hypothalamus is not essential to this stimulatory action of vasopressin. In their studies, however, the possibility of additional sites of action of vasopressin in the brain structures was not excluded.

Evidence for an extrapituitary site of action of vasopressin in stimulating pituitary ACTH release has also been provided. *Hedge* et al. [1966] showed that vasopressin injected into the anterior pituitary of dexamethasone-pretreated rats was less effective in elevating plasma corticosterone levels than that injected into the median eminence. They inferred that vasopressin acts on the median eminence rather than directly on the anterior pituitary. *Yasuda and Greer* [1976a] failed to find any significant stimulatory effect of vasopressin on ACTH release from cultured rat adenohypophyseal cells in vitro and suggested that pituitary ACTH release in vivo following vasopressin administration might not be due its direct action on the anterior pituitary.

Inhibitory effect of vasopressin on pituitary-adrenocortical secretory responses to stressful stimuli has been reported by some investigators. Adrenal ascorbic acid depletion in rats in response to epinephrine is suppressed by administration of Pituitrin [*Kimura*, 1954; *Arimura*, 1955/56], Pitressin [*Itoh and Arimura*, 1954; *Arimura*, 1955/56] and vasopressin but not oxytocin [*Itoh and Arimura*, 1954]. Adrenal ascorbic acid depletion in rats following exposure to cold (7 °C for 1 h) or histamine injection is also significantly suppressed by pretreatment with Pitressin [*Arimura*, 1955/56].

Oxytocin

Adrenal ascorbic acid depletion in rats following i.p. administration of Pitocin (a commercial preparation of oxytocic principle of posterior pituitary extract) has been observed [*Casentini* et al., 1959], a response completely abolished by hypophysectomy. *De Wied* et al. [1961] also noted adrenal ascorbic acid depletion in rats after i.p. injection of large doses of Pitocin or oxytocin. *Hilton* et al. [1960] failed to find any significant direct stimulatory effect of oxytocin on cortisol secretion of isolated perfused adrenal gland of hypophysectomized dogs.

C. Catecholamines, Acetylcholine and Serotonin

Catecholamines

A significant fall in adrenal ascorbic acid in rats following systemic administration of epinephrine has been reported by a large number of investigators [*Long and Fry*, 1945; *Long*, 1947; *Sayers and Sayers*, 1947; *Dury*, 1950; *Nasmyth*, 1950; *Ronzoni and Reichlin*, 1950; *Itoh and Arimura*, 1954; *Kimura*, 1954; *Briggs and Munson*, 1955; *George and Way*, 1955b; *Guillemin*, 1955; *Arimura*, 1955/56; *Ohler and Sevy*, 1956; *Olling and De Wied*, 1956; *Sevy* et al., 1957; *Van Peenen and Way*, 1957; *Casentini* et al., 1959; *Endröczi* et al., 1959; *Fisher and De Salva*, 1959; *Kitay* et al., 1959b; *Sapeika*, 1959; *Mitamura*, 1960b; *Giuliani* et al., 1961; *Itoh* et al., 1964; *Itoh and Yamamoto*, 1964]. Adrenal ascorbic acid response in rats to epinephrine is completely abolished by hypophysectomy [*Long and Fry*, 1945; *Long*, 1947] and by adenohypophysectomy [*Itoh and Yamamoto*, 1964], suggesting that it is mediated through the anterior pituitary. Adrenal cholesterol depletion in response to epinephrine in intact but not in hypophysectomized rats has also been observed [*Long and Fry*, 1945; *Long*, 1947].

Adrenal ascorbic acid in rats has been found to be significantly reduced by systemic administration of norepinephrine [*Nasmyth*, 1950; *Ronzoni and Reichlin*, 1950; *Guillemin*, 1955; *Olling and De Wied*, 1956; *Casentini* et al., 1959; *Sapeika*, 1959].

In ether-chloralose-anesthetized, eviscerated dogs and cats, whose adrenal venous plasma samples were assayed for adrenocortical hormone by the method of *Selye and Schenker* [1938], *Vogt* [1944] found an increase in adrenocortical secretion following i.v. infusion of epinephrine in doses which could be actually released from the adrenal medulla under some physiological conditions. *Endröczi* et al. [1958] noted an increase by 50–120% in adrenal 17-OHCS and corticosterone secretion in dogs anesthetized mostly with dial (in some cases with ether) following infusion of epinephrine at the rates of 3.0–3.6 µg/kg/min over 15 min. In contrast, *Katsuki* [1961] failed to find any demonstrable change in adrenal 17-OHCS secretion in dogs during and after i.v. infusion of epinephrine (20 or 50 µg/min over 10 min). *Gann and Egdahl* [1965] reported that in pentobarbital-anesthetized dogs an i.v. infusion of norepinephrine at a rate enough to elevate mean arterial pressure by 20 mm Hg did not significantly increase adrenal 17-OHCS secretion. The mean secretion rate of 17-OHCS before and after administration of norepinephrine was 1.3 ± 0.3 and 1.0 ± 0.3 µg/min, respectively.

In conscious dogs approximately a 2-fold increase in peripheral plasma 17-OHCS concentrations is induced by i.v. infusion over 60 min of 0.12 mg/kg epinephrine but not of 0.017 mg/kg norepinephrine [*Harwood and Mason*, 1956]. A significant elevation of plasma corticosterone levels in rats following systemic administration of epinephrine has been reported [*Fisher and De Salva*, 1959; *Itoh* et al., 1964; *Itoh and Yamamoto*, 1964; *Arimura* et al., 1967; *Tilders* et al., 1982]. *Van Loon* et al. [1971c] observed in unpublished experiments a marked increase in plasma corticosterone concentrations in rats after i.p. injection of norepinephrine. In man a significant fall in plasma cortisol levels during and after i.v. infusion of norepinephrine [*Wilcox* et al., 1975; *Few* et al., 1980] and a marked elevation 30–75 min after termination of norepinephrine infusion [*Few* et al., 1980] have been observed.

Van Loon et al. [1971a] showed that the adrenal 17-OHCS secretion in pentobarbital-anesthetized dogs in response to surgical stress (laparotomy) was not significantly altered by i.v. infusion of norepinephrine (33 μg/kg) over 30 min.

The depletion of adrenal ascorbic acid [*King and Thomas*, 1968] and elevation of plasma corticosterone levels [*King*, 1969] in rats following systemic administration of dopamine have been observed.

Van Loon et al. [1971a] reported that the adrenal 17-OHCS secretory response in pentobarbital-anesthetized dogs to surgical stress was not altered by systemic administration of dopamine (i.v. injection of 50 μg/kg followed by i.v. infusion of 40 μg/kg/min over 30 min, or i.v. infusion of 120 μg/kg/min over 30 min).

Acetylcholine

Adrenal ascorbic acid depletion in rats following systemic administration of acetylcholine has been reported [*Casentini* et al., 1957, 1959; *Giuliani* et al., 1961]. Hypophysectomy has been observed to abolish adrenal ascorbic acid depletion response in rats to acetylcholine [*Casentini* et al., 1957]. *Yamashita* et al. [1978] measured adrenal 17-OHCS secretion rates in pentobarbital-anesthetized dogs before and after i.v. injection of 1 mg/kg of acetylcholine. In intact dogs the adrenal 17-OHCS secretion was found to be markedly increased following administration of acetylcholine. It attained the peak levels at 10 min after the injection and returned to the basal levels mostly by 60 min. Adrenocortical secretory response in the dog to acetylcholine was abolished by hypophysectomy.

Serotonin

Systemic administration of serotonin induces adrenal ascorbic acid depletion in intact rats [*Georges*, 1957; *Moussaché and Pereira*, 1957; *Casentini* et al., 1959; *Fiore-Donati* et al., 1959; *Fischer* et al., 1959; *Sapeika*, 1959; *Giuliani* et al., 1961] but not in hypophysectomized rats [*Moussaché and Pereira*, 1957].

Direct adrenocortical secretory effect of serotonin at the adrenal level has been demonstrated. *Rosenkrantz* [1959] and *Rosenkrantz and Laferte* [1960] have shown that serotonin induces an increase in in vitro release of blue tetrazolium reactive substance from the incubated adrenal fragments of rabbits. *Verdesca* et al. [1961] observed that adrenal cortisol secretion of the perfused adrenal gland of the hypophysectomized dog was markedly increased by serotonin but not by 5-hydroxytryptophan, a serotonin precursor.

D. Adenosine 3',5'-Monophosphate

Evidence for a direct stimulating effect of adenosine 3',5'-monophosphate (cyclic AMP) on the adrenocortical cells has been accumulating.

Earp et al. [1970] observed in hypophysectomized rats a significant depletion of adrenal ascorbic acid after i.v. infusion of dibutyryl cyclic AMP.

The secretion rate of cortisol of the dog adrenal gland perfused in situ has been found to be increased by infusion of cyclic AMP into the arterial perfusion circuit [*Hilton* et al., 1960, 1961]. *Imura* et al. [1965] have shown a significant increase in adrenal corticosterone secretion in hypophysectomized rats and a significant elevation of plasma corticosterone levels in hypophysectomized or dexamethasone-treated rats following i.v. infusion of 20 mg cyclic AMP or dibutyryl cyclic AMP. *Gallant and Brownie* [1973] also noted a significant elevation of plasma corticosterone levels in hypophysectomized rats in response to dibutyryl cyclic AMP (7.0 mg/rat, i.v.).

In in vitro experiments with incubated rat adrenal slices or fragments [*Haynes* et al., 1959; *Hilton* et al., 1961; *Farese* et al., 1969; *Haksar and Péron*, 1971; *Vapaatalo* et al., 1972] and in superfusion experiments with decapsulated rat adrenal bisects [*Flack* et al., 1969; *Schulster* et al., 1970; *Flack and Ramwell*, 1972], corticosterone production by adrenocortical tissue has been found to be increased by cyclic AMP or dibutyryl cyclic AMP. Similar results have been obtained in incubation experiments with isolated rat adrenal cells [*Kitabchi and Sharma*, 1971; *Rivkin and Chasin*, 1971; *Sayers* et al., 1971; *Kolanowski* et al., 1974]. A direct stimulatory effect of cyclic AMP or dibutyryl cyclic AMP on corticosteroidogenesis has been demonstrated also in in vitro experiments with guinea pig [*Hilton* et al., 1961], beef [*Saruta and Kaplan*, 1972] and human [*Studzinski and Grant*, 1962; *Pearlmutter* et al., 1974a] adrenal tissue and trypsin (or collagenase-trypsin) dispersed guinea pig [*Kolanowski* et al., 1974], cat [*Rubin and Warner*, 1975] and dog [*Hirose* et al., 1979] adrenal cells.

E. Prostaglandins

Adrenal ascorbic acid and cholesterol depletion in pentobarbital-anesthetized rats following i.v. injection of prostaglandin E_1 (PGE$_1$)

has been observed [*Peng* et al., 1970]. It also has been shown that adrenal ascorbic acid response to PGE₁ is abolished by hypophysectomy and significantly inhibited by pretreatment with morphine and that plasma and adrenal corticosterone levels in pentobarbital-anesthetized rats are elevated 15 min after i.v. injection of PGE₁ [*Peng* et al., 1970].

In pentobarbital-chlorpromazine-treated rats a significant elevation of plasma corticosterone levels and a significant increase in corticosteroid production in vitro of excised adrenal glands are induced by i.v. injection of PGE₁ or PGE₂ but not PGF₁α or PGF₂α [*De Wied* et al., 1969]. In pentobarbital-anesthetized rats the above responses to PGE₁ are reduced by pretreatment with morphine and by neurohypophysectomy and are abolished by hypophysectomy [*De Wied* et al., 1969]. *Hedge* [1972] observed a significant increase in plasma concentration of corticosterone in rats following i.v. injection of PGE₁ or PGF₁α, an increase completely abolished by hypophysectomy. *Gallant and Brownie* [1973] failed to find any significant effect of a large dose of PGE₂ (200 μg/rat, i.p.) on plasma corticosterone levels in rats at 24 h after hypophysectomy. In rats hypophysectomized 3–6 h previously, however, *Flack* et al. [1969] showed significant elevations of plasma and adrenal corticosterone levels induced by 100 μg/100 g (i.p.) of PGE₂. Infusion of PGE₂ (1.6 μg/min) but not PGF₂α into the common carotid artery of fetal lambs [*Louis* et al., 1976] and injection of PGE₂ (50 μg) into the brachiocephalic artery of newborn lambs [*Challis* et al., 1978] result in a significant elevation of plasma cortisol levels. *Fichman* et al. [1972] have demonstrated a significant increase in peripheral plasma cortisol concentrations in man following i.v. infusion of PGA₁.

As to the direct effect of prostaglandins on corticosteroidogenesis in rats, the findings are still contradictory. In superfusion experiments with bisected decapsulated rat adrenals, the production of corticosterone has been found to be increased by PGE₂ [*Flack* et al., 1969; *Flack and Ramwell*, 1972], PGE₁ and PGF₂α [*Flack* et al., 1969]. *Spät and Józan* [1975] also noted that the production of corticosterone in vitro by rat capsular adrenal glands was stimulated by PGA₂. In contrast, *Lowry* et al. [1973] failed to find any stimulatory effect of PGE₂ on corticosterone production in vitro by trypsin-dispersed rat adrenal cells. *Matsuoka* et al. [1980] also observed that PGE₂, PGI₂ and PGF₂α were ineffective in stimulating corticosterone production in vitro by collagenase-dispersed rat adrenal decapsular cells.

Corticosterone and cortisol productions in vitro by beef adrenocortical fragments are significantly increased by PGE₁ and PGE₂ [*Saruta and Kaplan*, 1972]. Corticosteroid production in vitro of trypsin-dispersed cat adrenocortical cells has been observed to be significantly increased by PGE₂ [*Rubin and Warner*, 1975; *Warner and Rubin*, 1975; *Laychock and Rubin*, 1976; *Ellis* et al., 1978], PGE₁, PGF₁α [*Warner and Rubin*, 1975], PGI₂, PGH₂ or 6-keto-PGF₁α [*Ellis* et al., 1978]; PGI₂ is 100–1,000 times more effective than PGE₂ [*Ellis* et al., 1978]. A significant increase in cortisol and corticosterone productions in vitro of collagenase-trypsin-dispersed dog adrenal cells in response to PGE₂ has been noted [*Hirose* et al., 1979].

Honn and Chavin [1976] showed that the in vitro production of cortisol by human adrenal fragments was significantly increased by PGE₁ and PGE₂ and significantly depressed by PGF₁α and PGF₂α. *Honn and Chavin* [1977] reported that the production of cortisol in vitro by human adrenocortical tissue was significantly increased by 10–100 μg/ml of PGA₁, PGA₂, PGB₁ and PGB₂ and it was significantly depressed by 1 μg/ml of PGA₁ and PGA₂.

To evaluate the direct action of prostaglandin on the adrenal cortex, *Blair-West* et al. [1971] performed the close intraarterial infusion of PGE₁ to the autotransplanted adrenal gland of the sheep. Following the infusion of PGE₁ they observed elevations of mean adrenal cortisol and corticosterone secretion rates, which were however statistically not significant. In dexamethasone-treated immature dogs *Hirose* [1981] could demonstrate a significant increase in adrenal secretion of cortisol and corticosterone induced by infusion into the lumboadrenal artery of PGE₂.

The effect of indomethacin, an inhibitor of prostaglandin synthesis, on ACTH-induced adrenocortical secretion has been studied by some investigators. An elevation of plasma corticosterone levels in hypophysectomized rats in response to ACTH has been found to be reduced by pretreatment with indomethacin [*Gallant and Brownie*, 1973]. *Laychock and Rubin* [1976] demonstrated that an ACTH-induced increase in steroid release from the trypsin-dispersed cat adrenocortical cells in vitro was depressed by a high concentration (3×10^{-5} M) of indomethacin and potentiated by a low concentration (3×10^{-9} M) of indomethacin. *Honn and Chavin* [1976] observed that the stimulatory effect of ACTH on adrenocortical hormone production in vitro by human adrenal fragments was inhibited by indomethacin. In contrast,

Lowry et al. [1973] showed that the corticosteroid production in vitro by trypsin-dispersed rat adrenal cells in response to ACTH was not affected by indomethacin. A similar finding has been obtained in in vitro experiments with rat adrenal slice [*Vukoson* et al., 1976] and collagenase-dispersed rat adrenal decapsular cells [*Matsuoka* et al., 1980].

F. Neuropharmacologic Agents

Nicotine

Adrenal ascorbic acid [*Gray and Munson*, 1951; *Maren*, 1951; *Nowell*, 1959; *Anichkov* et al., 1960] and cholesterol [*Kershbaum* et al., 1968] depletions in rats following systemic administration of nicotine have been reported. Hypophysectomy abolishes adrenal ascorbic acid depletion in rats in response to nicotine [*Maren*, 1951].

Hume [1957] noted a transient increase in adrenal 17-OHCS secretion in dogs following i.v. injection of 0.1 mg/kg nicotine. In experiments with conscious dogs, *Anichkov* et al. [1960] drew the inferior vena caval blood samples from a point superior to the entrance of the adrenal vein through a thin polyethylene catheter inserted into the inferior vena cava. A marked elevation in plasma 17-OHCS levels of vena caval blood was observed following injection of nicotine (0.04–0.05 mg/kg) into the carotid body circulation. In pentobarbital-anesthetized dogs *Kershbaum* et al. [1968] also showed an increase in plasma 11-OHCS concentrations of inferior vena caval blood following i.v. injection of 0.9 mg/kg nicotine. *Suzuki* et al. [1973] evaluated directly the adrenal 17-OHCS secretion in conscious and anesthetized dogs in response to nicotine. 5 conscious dogs were injected i.v. with 0.1 mg/kg of nicotine. Adrenal 17-OHCS secretion was elevated from the resting secretion rates of 0.05–0.20 to 1.15–2.12 µg/kg/min, it attained the peak at 5 min after nicotine injection in 3 dogs and at 10 min in 2 dogs and it returned to the basal levels by 40 min (fig. 16). 5 pentobarbital-anesthetized dogs were also i.v. injected with 0.1 mg/kg nicotine. The adrenal 17-OHCS secretion rates were markedly elevated following nicotine injection (fig. 17); the adrenocortical secretory response to nicotine in pentobarbital-anesthetized dogs was not significantly different from that in conscious dogs. In hypophysectomized pentobarbital-anesthetized dogs, no significant increase in adrenal 17-OHCS secretion was observed following i.v. injection of 0.1 mg/kg nicotine. Thus,

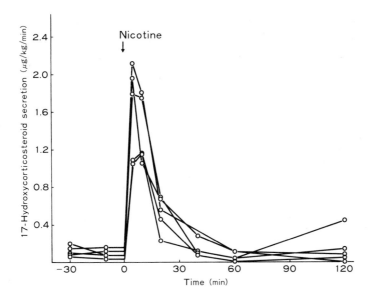

Fig. 16. Increase in adrenal 17-OHCS secretion in conscious dogs following i.v. injection of 0.1 mg/kg nicotine.

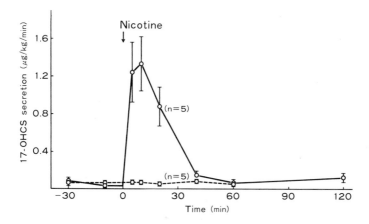

Fig. 17. Adrenocortical secretory response to nicotine (0.1 mg/kg, i.v.) in 5 intact (O——O) and 5 hypophysectomized (O- - -O) pentobarbital-anesthetized dogs. Data are presented as mean ± SEM. [Based on data from *Suzuki* et al., 1973.]

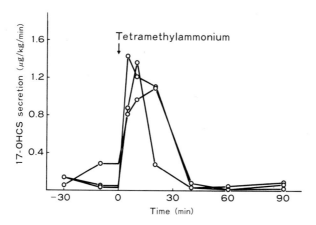

Fig. 18. Increase in adrenal 17-OHCS secretion rate in conscious dogs following i.v. injection of 1.0 mg/kg of tetramethylammonium bromide. [Based on data from *Suzuki* et al., 1965a.]

nicotine-induced adrenocortical secretion in the dog seems to be mediated through the pituitary gland. However, in vitro experiments by *Rubin and Warner* [1975] with trypsin-dispersed cat adrenocortical cells showed a possible direct stimulatory action of nicotine on adrenal corticosteroidogenesis.

An elevation of plasma corticosterone levels [*Balfour* et al., 1975; *Sithichoke and Marotta*, 1978; *Conte-Devolx* et al., 1981] and increases in plasma and adrenal 11-OHCS concentrations [*Kershbaum* et al., 1968] in rats following systemic administration of nicotine have been reported. *Hökfelt* [1961] observed a marked increase in peripheral plasma cortisol concentrations in man after smoking of non-filtered cigarettes. *Kershbaum* et al. [1968] also noted an elevation of peripheral plasma 11-OHCS levels in man after heavy cigarette smoking.

Tetramethylammonium

Suzuki et al. [1965a] studied the adrenocortical secretory response in the dog to tetramethylammonium. 3 conscious dogs were i.v. injected with 1.0 mg/kg of tetramethylammonium. The adrenal 17-OHCS secretion rate increased from the resting values of 0.03–0.29 to 1.10–1.43 µg/kg/min following the injection (fig. 18). It attained the

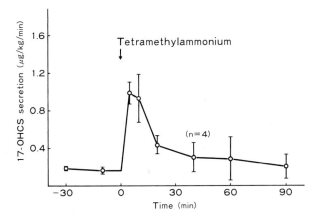

Fig. 19. Adrenal 17-OHCS secretion rates (mean ± SEM) in 4 pentobarbital-anesthetized dogs before and after i.v. injection of 1.0 mg/kg tetramethylammonium bromide. [Based on data from *Otsuka*, 1966.]

peak levels at 5–20 min after tetramethylammonium injection and returned to the basal levels by 40 min.

Otsuka [1966] showed that adrenal 17-OHCS secretion in the dog in response to tetramethylammonium was not suppressed by pentobarbital anesthesia (fig. 19).

Pilocarpine

Suzuki et al. [1975a] studied the effect of pilocarpine on adrenocortical secretion in conscious and anesthetized dogs. 3 conscious dogs were injected i.v. with 0.3 mg/kg pilocarpine. The adrenal 17-OHCS secretion rate increased from the basal levels of 0.07–0.17 μg/kg/min within 5 min after injection and attained the peak levels of 1.44–1.66 μg/kg/min at 5 min in 1 dog and at 40–60 min in 2 dogs (fig. 20). It returned to the basal levels by 120 min. In 4 pentobarbital-anesthetized dogs, the adrenal 17-OHCS secretion rate also increased markedly after i.v. injection of 0.3 mg/kg pilocarpine and returned to the basal values by 60 min (fig. 21). The extent of increase in adrenal 17-OHCS secretion following injection of pilocarpine in anesthetized dogs was not significantly different from that in conscious dogs. However, the duration of elevated adrenocortical secretion in anesthetized dogs was shorter than that in conscious dogs.

Adrenocortical secretory response to pilocarpine in pentobarbital-

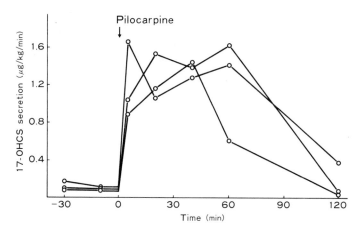

Fig. 20. Adrenal 17-OHCS secretion in conscious dogs in response to pilocarpine (i.v. injection of 0.3 mg/kg pilocarpine hydrochloride). [Based on data from *Suzuki* et al., 1975a.]

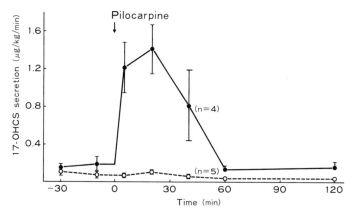

Fig. 21. Adrenocortical secretory response to pilocarpine (i.v. injection of 0.3 mg/kg pilocarpine hydrochloride) in 4 intact (●) and 5 hypophysectomized (○) pentobarbital-anesthetized dogs. Data are shown as mean ± SEM. [Based on data from *Suzuki* et al., 1975a.]

anesthetized dogs has been shown to be completely abolished by hypophysectomy [*Suzuki* et al., 1975a, b].

Eserine (Physostigmine)

Dordoni and Fortier [1950] reported a depletion of adrenal ascorbic acid in rats injected s.c. with eserine. Adrenocortical secretion in the

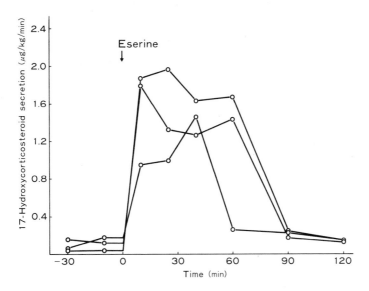

Fig. 22. Adrenal 17-OHCS secretion rates in conscious dogs before and after i.v. injection of 0.3 mg/kg eserine salicylate. [Based on data from *Suzuki* et al., 1964a.]

dog in response to eserine has been directly evaluated by *Suzuki* et al. [1964a]. In their study 3 conscious dogs were injected i.v. with 0.3 mg/kg of eserine. The adrenal 17-OHCS secretion rate increased from the resting values of 0.04–0.18 to 1.47–1.97 µg/kg/min (fig. 22). It attained the peak at 10–40 min after the injection of eserine and returned to the basal levels by 90 min. In pentobarbital-anesthetized dogs, *Otsuka* [1966] observed that adrenal 17-OHCS secretion was markedly elevated following i.v. injection of 0.3 mg/kg eserine (fig. 23), though the extent of elevation in anesthetized dogs was smaller than that observed by *Suzuki* et al. [1964a] in conscious dogs. The secretion rate attained the peak at 10–25 min after eserine injection and returned to the basal levels by 60 min. The finding suggests that the pituitary-adrenocortical secretory response in dogs to eserine might partially be suppressed by pentobarbital anesthesia. *Tanigawa* et al. [1968] showed that adrenal 17-OHCS secretion in pentobarbital-anesthetized dogs in response to eserine was abolished by hypophysectomy, thus suggesting that the stimulatory action of eserine on adrenocortical secretion in the dog may be mediated through ACTH release of the pituitary.

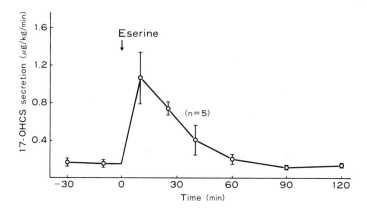

Fig. 23. Adrenocortical secretory response to eserine (i.v. injection of 0.3 mg/kg eserine salicylate) in 5 pentobarbital-anesthetized dogs. Data are presented as mean ± SEM. [Based on data from *Otsuka*, 1966.]

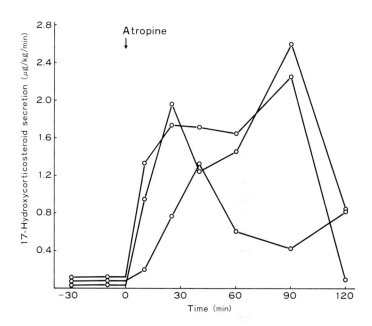

Fig. 24. Elevation of adrenal 17-OHCS secretion rate in conscious dogs following i.v. injection of 1.0 mg/kg atropine sulfate. [Based on data from *Suzuki* et al., 1964a.]

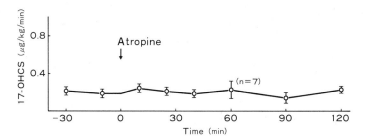

Fig. 25. Adrenocortical secretory response to atropine (i.v. injection of 1.0 mg/kg atropine sulfate) in 7 pentobarbital-anesthetized dogs. Data are presented as mean ± SEM. [Based on data from *Otsuka*, 1966.]

Harwood and Mason [1956] failed to find any elevation of peripheral plasma 17-OHCS levels in conscious dogs following a single i.v. injection of a small dose of eserine (0.05 mg/kg).

Atropine
Dordoni and Fortier [1950] observed a significant depletion of adrenal ascorbic acid in rats injected s.c. with atropine, a response completely abolished by hypophysectomy. *Guillemin* [1955] also noted a significant depletion of adrenal ascorbic acid in rats in response to atropine (s.c.).

Suzuki et al. [1964a] performed direct evaluation of adrenocortical secretion in conscious dogs in response to atropine. A marked increase in adrenal 17-OHCS secretion was observed without exception in 3 dogs injected i.v. with 1.0 mg/kg of atropine. As shown in figure 24, the adrenal 17-OHCS secretion rate, which was 0.02–0.12 μg/kg/min before atropine injection, increased and attained the peak (1.33–2.62 μg/kg/min) at 40 min after the injection of atropine in one dog and at 90 min in 2 dogs. At 120 min it still remained elevated in 2 dogs and returned to the pre-injection values in one dog.

In pentobarbital-anesthetized dogs, however, *Otsuka* [1966] failed to find any significant elevation in adrenal 17-OHCS secretion following i.v. injection of 1.0 mg/kg atropine (fig. 25). The finding indicates that pentobarbital anesthesia blocks the adrenocortical secretory response in the dog to atropine.

Plasma corticosterone levels in rats are elevated by i.p. injection of atropine [*Sithichoke and Marotta*, 1978]. However, a nocturnal rise in peripheral plasma 17-OHCS levels in cats has been found to be prevented by s.c. injection of atropine at 6 p.m. [*Krieger and Krieger*, 1967; *Krieger* et al., 1968].

Amphetamine

Adrenal ascorbic acid depletion in rats following systemic administration of amphetamine [*Nasmyth*, 1950] and hydroxyamphetamine [*Ohler and Sevy*, 1956] has been observed. However, *Bhattacharya and Marks* [1969a] found a significant depression of plasma and adrenal corticosterone concentrations in rats after i.p. injection of amphetamine. *Marantz* et al. [1976] showed a significant decrease in peripheral plasma cortisol concentrations in rhesus monkeys following i.v. injection of amphetamine.

Lorenzen and Ganong [1967] observed that adrenal 17-OHCS secretion in pentobarbital-anesthetized dogs in response to surgical stress was significantly depressed 30 min after i.v. administration of 1.5–4.0 mg/kg of amphetamine or 3–5 mg/kg of methamphetamine.

α-Methyl-p-Tyrosine (Tyrosine Hydroxylase Inhibitor)

A rise of plasma corticosterone levels in rats has been observed at 2 and 4 h [*Carr and Moore*, 1968], at 3–9 h [*Scapagnini* et al., 1975] and at 9 h [*Scapagnini* et al., 1970, 1971b; *Van Loon* et al., 1971c] after i.p. injection of α-methyl-p-tyrosine (α-MpT). *Kaplanski* et al. [1972] showed that corticosteroid production in vitro of excised incubated adrenal glands and plasma corticosterone levels in rats were significantly elevated 9 h after i.p. administration of α-MpT and these responses were abolished by pentobarbital anesthesia. *Krieger and Rizzo* [1969] noted an elevation of peripheral plasma 17-OHCS levels in conscious cats in response to α-MpT.

Apomorphine (Dopamine Receptor Agonist)

Peripheral plasma concentration of cortisol in conscious rhesus monkeys significantly increases following i.v. infusion of a large dose of apomorphine [*Chambers and Brown*, 1976]. Peripheral plasma corticosteroid levels in conscious dogs are significantly elevated by i.v. infusion of apomorphine [*Holland* et al., 1978].

Pargyline (Monoamine Oxidase Inhibitor)

An increase in plasma corticosterone in rats in response to surgical stress (exposure to ether vapor followed by laparotomy) has been found to be significantly suppressed by pretreatment with pargyline (i.p.) [*Bhattacharya and Marks*, 1969a]. *Marotta* et al. [1976] observed in pargyline (i.p.)-treated rats a significant elevation in plasma corticosterone levels and a significant depression of plasma corticosterone response to hypoxic stress (exposure to 10% O_2 + 90% N_2) and to hypercapnic stress (exposure to 10% CO_2 + 20% O_2 + 70% N_2).

α-Ethyltryptamine (Monase, Monoamine Oxidase Inhibitor)

Adrenal 17-OHCS secretion in pentobarbital-anesthetized dogs but not in ether-anesthetized dogs has been found to be transiently but significantly depressed by i.v. administration of α-ethyltryptamine (5 mg/kg) [*Tullner and Hertz*, 1964]. *Lorenzen and Ganong* [1967] found an inhibitory effect of α-ethyltryptamine (10 mg/kg, i.v.) and related drugs such as α-methyltryptamine (10 mg/kg, i.v.) on adrenal 17-OHCS secretion in pentobarbital-anesthetized dogs in response to surgical stress. *Ganong* et al. [1967] showed that the inhibitory effect of α-ethyltryptamine on adrenal 17-OHCS secretion in pentobarbital-anesthetized dogs in response to surgical stress was prevented by hemorrhage and suggested that this inhibitory effect of α-ethyltryptamine might be attributed to its pressor effect. *Ganong* et al. [1965] found that an increase in adrenal 17-OHCS secretion in pentobarbital-anesthetized dogs induced by electrical stimulation of the femoral nerve, caudal medulla oblongata, bundle of Schütz or mammillary peduncle but not the ventral hypothalamus or median eminence is blocked by pretreatment with α-ethyltryptamine. The findings suggest that α-ethyltryptamine might block the conduction from the afferent fibers to the neurosecretory CRF-releasing cells in the ventral hypothalamus.

FLA-63 (Dopamine-β-Hydroxylase Inhibitor)

Plasma corticosterone levels in rats were significantly elevated by i.p. administration of FLA-63 [*Scapagnini* et al., 1971b; *Marotta* et al., 1976].

G. Tranquilizers

Chlorpromazine

Although a systemic administration of small doses (1.0–2.5 mg/kg) of chlorpromazine in rats has been shown to have no adrenal ascorbic acid-depleting effect [*Nasmyth*, 1955; *Nagata*, 1957; *Mahfouz and Ezz*, 1958; *Ashford and Shapero*, 1962], most investigators [*Holzbauer and Vogt*, 1954; *Nasmyth*, 1955; *Kovács* et al., 1956; *Nagata*, 1957; *Sapeika*, 1959; *Ashford and Shapero*, 1962; *Smith* et al., 1963; *Degonskii*, 1977] have observed a significant depletion of adrenal ascorbic acid in rats induced by moderate or large doses of chlorpromazine. In contrast to most investigators, *Sevy* et al. [1957] failed to find any significant depletion in rats following s.c. injection of a large dose (20 mg/kg) of chlorpromazine. Chlorpromazine-induced depletion of adrenal ascorbic acid in rats is abolished by daily doses of 2.8 mg/kg (s.c.) of chlorpromazine for 5 days [*Ashford and Shapero*, 1962] and by hypophysectomy [*Nagata*, 1957; *Ashford and Shapero*, 1962; *Smith* et al., 1963].

Egdahl and Richards [1956a] examined the effect of chlorpromazine on adrenal 17-OHCS secretion in conscious and pentobarbital-anesthetized dogs. The adrenal 17-OHCS secretion rate was found to be increased in about 50% of dogs injected i.v. with 1.3–3.3 mg/kg chlorpromazine and in all dogs injected i.v. with 3.5–10.0 mg/kg chlorpromazine; the maximum secretion was mostly attained 6–12 min after injection of chlorpromazine in the former dogs and 3–9 min in the latter dogs. The adrenocortical secretory response in dogs to chlorpromazine was not affected by pentobarbital anesthesia, but it was completely abolished by hypophysectomy. *Bohus and Endröczi* [1964a] showed a significant increase in the adrenal corticosterone secretion rate in rats following s.c. injection of chlorpromazine (2–10 mg/kg).

Elevations of plasma [*Smith* et al., 1963; *Bhattacharya and Marks*, 1969b] and adrenal [*Bhattacharya and Marks*, 1969b] corticosterone levels in rats and elevation of plasma 17-OHCS in guinea pigs [*Sasaki*, 1963] after administration of chlorpromazine have been reported. The plasma corticosterone response in rats to chlorpromazine is abolished by hypophysectomy [*Smith* et al., 1963]. *Harwood and Mason* [1957] evaluated the peripheral plasma 17-OHCS response in monkeys (*Macaca mulatta*) to chlorpromazine (0.5–5.0 mg/kg, i.v.); graded plasma 17-OHCS elevations were found to be induced by 0.5 and 2.5 mg/kg of this drug, whereas 5.0 mg/kg were less effective than 2.5 mg/kg in elevating plasma 17-OHCS levels.

Studies of the effect of chlorpromazine on adrenocortical secretory response to various stressful stimuli have revealed conflicting results.

Nasmyth [1955] has shown that adrenal ascorbic acid depletion in rats in response to a small dose (10 mg/kg, s.c.) but not a large dose

(100 mg/kg, s.c.) of histamine is blocked by pretreatment with chlorpromazine. Pretreatment with chlorpromazine in rats has been reported to suppress adrenal ascorbic acid depletion induced by i.v. injection of 0.9% saline or 10% formalin solution [*Cheymol* et al., 1954] and the depletion in response to surgical stress [*Hamburger*, 1955; *Sevy* et al., 1957], epinephrine [*Olling and De Wied*, 1956; *Sevy* et al., 1957; *Van Peenen and Way*, 1957], norepinephrine, histamine [*Olling and De Wied*, 1956], morphine [*Van Peenen and Way*, 1957], ether anesthesia followed by bleeding, exposure to cold or heat [*Mahfouz and Ezz*, 1958] and exposure to vertical spinning [*Pohujani* et al., 1969]. In contrast, it has also been reported that pretreatment with chlorpromazine is ineffective in suppressing adrenal ascorbic acid depletion in rats following i.p. injection of hypertonic saline solution [*Kovács* et al., 1956], 1 N-acetic acid solution [*Nagata*, 1957] and the depletion in response to surgical stress [*Holzbauer and Vogt*, 1954; *Nasmyth*, 1955; *Nagata*, 1957], epinephrine [*Holzbauer and Vogt*, 1954; *Nasmyth*, 1955], aspirin, histamine [*Van Peenen and Way*, 1957], ether [*Ashford and Shapero*, 1962] and exposure to heat [*Degonskii*, 1977].

Betz and Ganong [1963] studied the effect of chlorpromazine on adrenal 17-OHCS secretion response in dogs to surgical stress and peripheral plasma 17-OHCS response in dogs to immobilization stress. They observed that adrenal 17-OHCS secretion in pentobarbital-anesthetized dogs in response to surgical stress was not significantly altered by the administration of 5 mg/kg chlorpromazine (a single i.v. dose of 2.5 mg/kg followed by an i.v. infusion of 2.5 mg/kg over 30 min). In conscious dogs pretreated with chlorpromazine in daily doses of 30 mg/kg for 3 days, a rise of plasma 17-OHCS levels after a 2-hour immobilization was not different from that in control animals without chlorpromazine pretreatment.

Bohus and Endröczi [1964a] showed that the adrenal secretion of corticosterone in rats in response to electroshock, immobilization or s.c. injection of 5% formalin solution was not suppressed by pretreatment with chlorpromazine (0.5–10 mg/kg, s.c.).

Lammers and De Wied [1963] observed that the adrenocortical secretory response in rats, as measured by production of corticosteroids in vitro of excised adrenal glands, to various stressful stimuli, such as exposure to a strange environment or auditory stimulation, was not depressed by pretreatment with a small dose (2 mg/kg) of chlorpromazine. However, the same investigators [*Lammers and De Wied*, 1964] demonstrated that the pituitary-adrenocortical secretory response (production of corticosteroids in vitro by excised adrenal glands) in pentobarbital-anesthetized rats to nicotine, histamine and surgical stress was depressed by pretreatment with a large dose (20 mg/kg) of chlorpromazine.

An elevation of plasma corticosterone levels in rats following exposure to cold (4 °C) for 2 h has been shown to be abolished by pretreatment with 3 doses of 20 mg/kg chlorpromazine (i.p.) at 8-hour intervals [*Smith* et al., 1963]. In man *Christy* et al. [1957] observed that an insulin-induced elevation of peripheral plasma 17-OHCS levels was blocked by pretreatment with 100–300 mg of chlorpromazine (oral administration of 50–150 mg twice with a 2-hour interval).

Reserpine

Systemic injection of reserpine has been shown to reduce adrenal ascorbic acid in rats [*Wells* et al., 1956; *Van Peenen and Way*, 1957; *Kitay* et al., 1959b; *Saffran and Vogt*, 1960; *Khazan* et al., 1961; *Maickel* et al., 1961; *Ashfold and Shapero*, 1962; *Gaunt* et al., 1962; *Martel* et al., 1962; *Westermann* et al., 1962], though small doses of reserpine have been noted to be ineffective in rats in depleting adrenal ascorbic acid [*Mahfouz and Ezz*, 1958; *Westermann* et al., 1962]. Adrenal ascorbic acid depletion in rats in response to reserpine is abolished by hypophysectomy [*Maickel* et al., 1961]; it is significantly depressed or completely abolished by daily pretreatment with this drug for several days [*Wells* et al., 1956; *Khazan* et al., 1961; *Ashford and Shapero*, 1962].

Egdahl et al. [1956] examined the effect of reserpine on adrenal 17-OHCS secretion in conscious dogs. In their study the animals were i.v. injected with 0.12–0.33 mg/kg of reserpine and were drowsy during the 3-hour period after the injection. A marked increase in adrenal 17-OHCS secretion was observed following administration of reserpine without exception.

Elevations of plasma [*Maickel* et al., 1961; *Eechaute* et al., 1962b; *Martel* et al., 1962; *Montanari and Stockham*, 1962; *Westermann* et al., 1962; *Carr and Moore*, 1968; *Bhattacharya and Marks*, 1969a, b; *McKinney* et al., 1971; *Lengvári and Halász*, 1972; *Scapagnini* et al., 1976] and adrenal [*Montanari and Stockham*, 1962; *Bhattacharya and Marks*, 1969a, b] corticosterone levels in rats following systemic administration of reserpine have been reported. Plasma [*Maickel* et al.,

1961; *Montanari and Stockham*, 1962] and adrenal [*Montanari and Stockham*, 1962] corticosterone responses to reserpine are completely blocked in hypophysectomized rats. *Eechaute* et al. [1962b] noted an increase in adrenocortical secretory activity, as measured by corticosteroid production in vitro of excised adrenal glands, in rats after injection of reserpine.

In monkeys (*Macaca mulatta*) an elevation of plasma 17-OHCS levels is induced by a single i.v. injection of reserpine [*Harwood and Mason*, 1957].

The site of action of reserpine on the pituitary-adrenocortical secretory activity has been suggested to be localized in some brain structures outside the medial basal hypothalamus [*Lengvári and Halász*, 1972].

The effect of pretreatment with reserpine on adrenocortical secretory response to various stimuli has been the subject of controversy. Adrenal ascorbic acid depletion in rats in response to histamine [*Wells* et al., 1956], ether anesthesia [*Wells* et al., 1956; *Kitay* et al., 1959b; *Montanari and Stockham*, 1962], epinephrine, morphine [*Van Peenen and Way*, 1957], exposure to cold [*Maickel* et al., 1961; *Mahfouz and Ezz*, 1958], exposure to heat [*Mahfouz and Ezz*, 1958], auditory stimulation [*Nowell*, 1959] and vertical spinning [*Pohujani* et al., 1969], and an elevation of plasma corticosterone levels in rats in response to cold exposure [*Maickel* et al., 1961] are significantly depressed by pretreatment with reserpine. In contrast, it has been noted that adrenal ascorbic acid depletion in rats induced by i.v. infusion of hypertonic saline solution or by cold exposure [*Nowell*, 1959], an elevation of plasma corticosterone levels in rats following exposure to cold [*Eechaute* et al., 1962b] or ether vapor [*Montanari and Stockham*, 1962] and an increase in adrenocortical secretory activity (corticosteroid production in vitro of excised adrenal glands) in rats following cold exposure [*Eechaute* et al., 1962b] are not impaired by pretreatment with reserpine.

H. Salicylate

Depletion of adrenal ascorbic acid in rats following systemic administration of sodium salicylate [*Blanchard* et al., 1950; *Hetzel and Hine*, 1951; *Cronheim* et al., 1952; *Coste* et al., 1953; *Giuliani* et al., 1961; *Wexler*, 1963], salicylic acid [*Petersen and Weidmann*, 1955;

Weidmann, 1955], acetylsalicylic acid (aspirin) [*Petersen and Weidmann*, 1955; *George and Way*, 1957; *Van Peenen and Way*, 1957; *Way* et al., 1962] and calcium acetylsalicylate [*Hetzel and Hine*, 1951] has been reported. Studies of the effect of hypophysectomy on adrenal ascorbic acid depletion in rats in response to salicylate or salicylic acid, however, revealed conflicting results. *Coste* et al. [1953], *Petersen and Weidmann* [1955] and *Weidmann* [1955] observed that adrenal ascorbic acid depletion in response to sodium salicylate or salicylic acid was not abolished, though it was partially suppressed, by hypophysectomy. In contrast, it was noted that adrenal ascorbic acid depletion was not induced in hypophysectomized rats by systemic administration of sodium salicylate [*Hetzel and Hine*, 1951; *Van Cauwenberge*, 1951; *Cronheim* et al., 1952] and aspirin [*Way* et al., 1962].

I. Bacterial Products

Adrenal ascorbic acid in rats has been found to be significantly depleted by i.p. injection of typhoid vaccine [*Sayers and Sayers*, 1947] and Piromen (bacterial polysaccharide of *Pseudomonas* origin) [*Wexler* et al., 1957, 1958; *Wexler*, 1963]. Adrenal ascorbic acid and cholesterol depletions in rats in response to Piromen are completely blocked by hypophysectomy [*Wexler* et al., 1957].

Melby et al. [1957] reported that adrenal 17-OHCS secretion in anesthetized dogs was markedly increased for 2 h following i.v. injection of endotoxin of *Brucella melitensis* (5 mg/kg) and this adrenocortical response was abolished by hypophysectomy. In conscious dogs, *Egdahl* [1959] observed that i.v. administration of 0.01–0.5 mg of purified *E. coli* endotoxin induced an elevation of body temperature over 1 °C and a marked increase in adrenal 17-OHCS secretion. The adrenocortical secretory response to endotoxin was not impaired, but the adrenal medullary secretory response was completely abolished, by the spinal cord transection at the level of C_7.

In dogs injected with 0.01 mg of purified *E. coli* endotoxin on consecutive days *Egdahl* et al. [1959] observed a significant increase in adrenal 17-OHCS secretion with a body temperature elevation of at least 1.5 °C on the first day of endotoxin administration and that without any elevation of body temperature on days 2 and 3. *Melby* et al. [1960] also evaluated adrenal cortisol secretion in pentobarbital-anesthetized dogs in response to bacterial endotoxin. Following i.v. injection of 1.0 mg/kg of purified *E. coli* endotoxin adrenal cortisol (17-OHCS) secretion was elevated. It increased from the basal value (less than 1 µg/min) to a mean value of 8.9 µg/min with a range of 4.1–15.7 µg/min at 30 min and to 9.6 (6.0–17.4) µg/min at 120 min. After i.v. injection of 0.002 mg of purified *Salmonella abortus equi* endotoxin (Pyrexal) it was elevated at 60 min to 8.9 (4.0–11.2) µg/min. The adrenocortical secretory response in dogs to endotoxin was found to be abolished by hypophysectomy.

Chowers et al. [1966] found a rise in peripheral plasma levels of cortisol prior to an appreciable body temperature elevation in dogs injected i.v. with bacterial pyrogen.

An elevation of plasma corticosterone levels in rats has been observed following i.p. injection of typhoid-paratyphoid vaccine and Piromen [*Moberg*, 1971]. *Yasuda and Greer* [1978] showed an elevation of plasma corticosterone levels in rats following i.p. injection of *E. coli* endotoxin, a response completely abolished by hypophysectomy.

An elevation of peripheral plasma 17-OHCS (or cortisol) levels in man following i.v. administration of typhoid vaccine [*Christy* et al., 1956], Piromen [*Bliss* et al., 1954; *McDonald* et al., 1956; *Kimball* et al., 1968; *Miller and Moses*, 1968], bacterial pyrogen (Organon) [*Shuster and Williams*, 1961; *Takebe* et al., 1966; *Jenkins*, 1968] and *Salmonella abortus equi* endotoxin [*Kimball* et al., 1968] has been reported.

Chapter 4. Responses of Adrenocortical Cells to Adrenocorticotropic Hormone

*A. Intracellular and Membrane Potential Events of
Adrenocortical Cells in Response to Adrenocorticotropic Hormone*

*Adenosine 3′,5′-Monophosphate, an Intracellular Mediator of the
Stimulatory Action of ACTH on Adrenocortical Cells*

The first event of the effect of ACTH on adrenocortical cells is the binding of ACTH to the ACTH receptors without entering adrenocortical cells and this binding to the receptor sites results in adenylate cyclase activation in adrenocortical cells [*Lefkowitz* et al., 1970a,b]. Adrenal adenylate cyclase activated by ACTH promotes the intracellular production of adenosine 3′,5′-monophosphate (cyclic AMP) which in turn increases adrenal phosphorylase activity [*Haynes, 1958*]. Activated adrenal phosphorylase accelerates the conversion of glycogen, via glucose-1-phosphate, to glucose-6-phosphate, which is then dehydrogenated; reduced triphosphopyridine nucleotide (TPNH), generated in the process of glucose-6-phosphate dehydrogenation, provides the energy necessary for some steps of corticosteroidogenesis [*Haynes and Berthet,* 1957].

Haynes [1958] has proposed a theory that cyclic AMP serves as a mediator through which ACTH stimulates corticosteroidogenesis. This theory has subsequently been supported by the findings that the corticosteroid production in adrenocortical cells is stimulated by cyclic AMP (cf. chapter 3. D) and that the production of cyclic AMP in vitro in adrenocortical cells [*Grahame-Smith* et al., 1967; *Beall and Sayers,* 1972; *Mackie and Schulster,* 1973; *Moyle* et al., 1973] and the release of cyclic AMP in vivo into the adrenal vein blood [*Peytremann* et al., 1973; *Espiner* et al., 1974; *Jarrett* et al., 1976] are increased by ACTH.

*Role of Calcium in the Stimulatory Action of ACTH on
Adrenocortical Cells*

In vitro incubation experiments with rat adrenal fragments [*Birmingham* et al., 1953] and isolated rat adrenal cells [*Haksar and Péron,*

1972; *Sayers* et al., 1972] have shown that removal of calcium from incubation medium abolishes or markedly depresses the ACTH-stimulated corticosterone production. Experiments with perfused cat adrenal gland have demonstrated that adrenal corticosteroid secretion in
response to ACTH is markedly depressed by removal of calcium from
perfusion fluid [*Jaanus* et al., 1970]. Thus, it is suggested that calcium
plays an essential role in the mechanism of corticosteroid release.

    In adrenal medullary cells acetylcholine, the transmitter of secre-
tory nerve impulses to the adrenal medullary cells, increases the permeability of the cell membrane to various ions and thus promotes inward movement of extracellular calcium through the cell membrane
[*Douglas and Rubin*, 1961; *Douglas and Poisner*, 1962]. After penetra-
tion, calcium acts as a trigger to stimulate the movement of secretory
granules in adrenomedullary cells toward the cell membrane and the
fusion of the former to the latter, and thus to promote the release of
catecholamines contained in the granules.

    In adrenocortical cells, however, *Jaanus and Rubin* [1971] failed to
find any net inward movement of calcium in response to ACTH. These
investigators and *Rubin* et al. [1972] proposed a theory that ACTH
might induce a translocation of calcium in adrenocortical cells from a
rapidly exchangeable pool to a less readily exchangeable pool.

    Lefkowitz et al. [1970a] studied the effect of calcium on the ACTH
binding at the receptor sites and the activation of adenylate cyclase in
incubation experiments with preparations from ACTH-sensitive
mouse adrenal tumor. They observed that removal of calcium from incubation medium by EGTA markedly depressed the ACTH-induced
activation of adrenal adenylate cyclase but not the binding of ACTH
and suggested that calcium was essential for the process between
ACTH binding at receptor sites and activation of adenylate cyclase.
Sayers et al. [1972] stated that calcium might exert an effect on the
ACTH receptor-adenylate cyclase complex in amplifying the signal
generated by ACTH.

Membrane Potential Events of Adrenocortical Cell in
Response to ACTH
Matthews [1967] and *Matthews and Saffran* [1967] have measured
the resting membrane potential of incubated adrenocortical cells in vitro using glass microelectrodes. Mean values of adrenocortical membrane potential in the rat, rabbit and kitten are 70.5, 66.2 and 71.4 mV,

respectively. As expected, this membrane potential largely depends on the gradient between intra- and extracellular potassium concentrations; the cell membrane is depolarized by raising extracellular potassium concentration. It has been observed that ACTH does not alter the adrenocortical membrane potential though it stimulates corticosteroid production, suggesting that there is no direct correlation between the former and the latter.

Later, *Matthews and Saffran* [1968, 1973] have found ACTH-induced action potential-like changes (amplitude: 10–60 mV) in neonatal-rabbit adrenocortical cells incubated in a K^+-free medium. Initiation of the potential changes is not blocked by tetrodotoxin (10^{-6} g/ml), indicating that it is not Na^+-dependent.

Lymangrover [1980] showed a rapid and transient depolarization, ranged from 2.8 to 8.0 mV, of transmembrane potential in response to ACTH in mouse adrenocortical cells superfused with Krebs-Henseleit-glucose solution. It is suggested that this membrane potential change might be due to an alteration of membrane permeability to some specific ions, e.g. Ca^{++}. *Lymangrover* et al. [1982] extended this study and found that the ACTH-induced depolarization of membrane potential of superfused mouse adrenocortical cells was completely blocked by Ca^{++}-blocking agents such as $CoCl_2$ and verapamil, whereas ACTH-stimulated corticosteroidogenesis was not completely blocked, though it was significantly inhibited, by $CoCl_2$. From these and other findings they proposed a theory that two mechanisms were involved in the stimulatory action of ACTH on corticosteroidogenesis; (1) ACTH increases intracellular production of adrenal cyclic AMP which directly (and indirectly via alteration of ionic contents) activates intracellular enzymic systems for corticosteroidogenesis, and (2) ACTH alters directly the cell membrane permeability to specific ions which then stimulate directly or indirectly the enzymic systems involved in corticosteroidogenesis.

B. Factors Affecting the Responsiveness of the Adrenal Cortex to Adrenocorticotropic Hormone

Corticosteroids – Direct Inhibitory Effect of Corticosteroids at the Adrenal Level on Corticosteroidogenesis
Direct feedback inhibition at the adrenal level by corticosteroids

of ACTH-stimulated adrenal steroidogenesis has been studied by a number of investigators. Production of corticosteroids [*Birmingham and Kurlents,* 1958] and corticosterone [*Péron* et al., 1960; *Fukui* et al., 1961; *Fekete and Görög,* 1963] in vitro by incubated rat adrenal fragments in response to ACTH is significantly depressed by corticosterone, cortisol or other synthetic corticosteroids such as prednisolone and dexamethasone added into the incubation medium. In vitro experiments by *Abe and Hirose* [1974] with trypsin-dispersed rat adrenal cells demonstrated a significant decrease in ACTH-stimulated corticosterone production following addition of corticosterone (1-2 µg) into the incubation medium (1 ml). Similar observations were reported by *Carsia and Malamed* [1979]. They observed that corticosterone production by trypsin-dispersed rat adrenal cells in vitro in the presence of ACTH was significantly inhibited by 0.5-2.0 µg/ml of corticosterone or cortisol but not by 1.0 µg/ml of aldosterone. They also showed that a 22% suppression of ACTH-induced corticosteroidogenesis in vitro in trypsin-dispersed beef adrenocortical cells was produced by 0.25 µg/ml of exogenous cortisol and a 33% inhibition by 1.0 µg/ml cortisol. Inhibition by cortisol of binding of ACTH to adrenocortical plasma membrane, which has been demonstrated in in vitro experiments by *Latner* et al. [1977], may be a possible mechanism of direct inhibitory effect of corticosteroids on ACTH-induced corticosteroid production.

In contrast to the above investigations, *Vinson* [1966] failed in in vitro experiments with minced rat adrenal tissue to find any significant inhibition by corticosterone added in the incubation medium in amounts of 3-10 µg/ml on the rate of conversion of [4-¹⁴C]progesterone to corticosterone. In this study, however, an inhibition by exogenous corticosterone on resting rather than ACTH-induced adrenal steroidogenesis was evaluated.

In experiments with perfused dog adrenal glands, *Black* et al. [1961] showed that adrenal secretion of cortisol and corticosterone in response to ACTH (10 mU/min) was markedly depressed by perfusion with 1-4 µg/ml (10-40 µg/min) of cortisol or synthetic corticosteroids, such as prednisolone or dexamethasone, and that adrenal secretion of corticosterone, but not cortisol, was also suppressed by 2 µg/ml (20 µg/min) of corticosterone. *Hill and Singer* [1968] have shown that adrenal corticosterone secretion in hypophysectomized rats into the adrenal vein in response to ACTH is reduced by about 22% when peripheral plasma levels of corticosterone are elevated by s.c. injection of corticosterone to the levels which are usually observed in intact rats subjected to surgical stress, i.e. about 40 μg/100 ml.

Adrenal Blood Flow

Porter and Klaiber [1964] have noted that adrenal cortical secretion in rats in response to ACTH is unaffected by changes in adrenal blood flow. However, the same investigators showed in their subsequent study [*Porter and Klaiber,* 1965] that adrenal corticosterone secretion in hypophysectomized rats infused i.v. with ACTH was significantly affected by variations in adrenal blood flow. *Urquhart* [1965] observed in experiments with perfused dog adrenal gland that the cortisol secretion rate did not correlate with adrenal blood flow in the presence of high blood concentrations (or in the absence) of ACTH, but it was significantly influenced by changes in adrenal blood flow when blood concentrations of ACTH were low or intermediate. Similar findings were obtained by *L'Age* et al. [1970] in experiments on intact conscious dogs. When the animals were i.v. infused with 2 mU/min of ACTH, an increase in adrenal blood flow resulted in an increase in cortisol secretion rate. During i.v. infusion of 25 mU/min of ACTH, however, the secretion rate of cortisol was not increased by an increased adrenal blood flow. *Shibata* et al. [1972] observed that adrenal 17-OHCS secretion in hypophysectomized pentobarbital-anesthetized dogs infused i.v. with large doses of ACTH (100–132 mU/min) was depressed when adrenal blood flow was extremely decreased by hemorrhage.

Hypothermia

Egdahl et al. [1955] have demonstrated that adrenal 17-OHCS secretion in dogs subjected to surgical stress with or without i.v. infusion of ACTH is markedly depressed when severe hypothermia (21–30 °C) is induced by cooling an external vascular shunt or by ice water immersion. Localized adrenal cooling in dogs with systemic normothermia was also found to be effective in depressing adrenal 17-OHCS secretion. *Ganong* et al. [1955a] also observed that adrenal 17-OHCS secretion in pentobarbital-ether-anesthetized dogs in response to endogenous or exogenous ACTH was decreased by severe hypothermia (24 °C) induced by immersion in an ice-water bath.

The above experimental results may be interpreted to indicate a direct suppressive effect of hypothermia at the adrenal level on the secretory responsiveness of the adrenal cortex to ACTH.

Localized Adrenal X-Irradiation

Yamashita and Shimizu [1972] have evaluated the responsiveness of the adrenal cortex to exogenous ACTH in pentobarbital-anesthetized dogs 2–14 days after localized adrenal irradiation with 2,000 R of X-ray. An increase in adrenal 17-OHCS secretion induced by i.v. injection of ACTH (5 mU/kg) in X-irradiated dogs was sustained up to 30 min and in some cases even up to 60 min after the injection, whereas the elevation of 17-OHCS secretion in non-irradiated dogs was observed at 15 min but not 30 and 60 min after the ACTH injection. The experimental results show an increased adrenocortical secretory responsiveness to ACTH in dogs 2–14 days after localized adrenal X-irradiation.

Chapter 5. Circadian Rhythm of Adrenocortical Secretory Activity

Pincus [1943] was the first to suggest the presence of a circadian (approximately 24-hour) variation of adrenocortical secretory activity. He has shown a diurnal variation of urinary excretion of 17-ketosteroids in man with a peak occurring during morning hours. Since then, the circadian periodicity of adrenocortical secretory activity has been extensively studied by a large number of investigators and there is now a voluminous literature in support of its existence in man and in various animal species.

A. Studies in Man

Many investigators have demonstrated the circadian periodicity of plasma concentration of 17-OHCS (or 11-OHCS or cortisol) in man (table IV).

In normal subjects, the peripheral plasma concentration of 17-OHCS shows a sharp rise to the peak levels which are attained during early morning between 0500 and 0900 hours, mostly between 0600 and 0800 hours, i.e. around the time of awakening. Then it shows a progressive decline and reaches the lowest levels during late evening and midnight, i.e. around the time of the onset of sleeping.

It has been reported that the circadian variations of peripheral plasma 17-OHCS levels in nurses and watchmen who have been working at night for more than 6 months and in completely blind subjects are found to be the same as those in normal subjects [*Migeon* et al., 1956]. However, *Perkoff* et al. [1959] observed that the circadian rhythm in peripheral plasma 17-OHCS levels was reversed in normal subjects when subjected to reversal of their sleep-wake schedules (an 8-hour shift of sleeping time). *Orth* et al. [1967] reported that after several days of adaptation to new sleep-wake patterns of 12, 19 or 33 h per cycle, one third of each cycle was spent in sleep, the circadian rhythm of

Table IV. Reports of circadian variations of peripheral plasma levels of 17- or 11-hydroxycorticosteroids or cortisol in man

Bliss et al. [1953]	*Krieger* et al. [1969]	*Okuyama* et al. [1977]
Sandberg et al. [1954]	*Orth and Island* [1969]	*Milcu* et al. [1978]
Doe et al. [1956]	*Hamanaka* et al. [1970]	*Reschini and Giustina*
Migeon et al. [1956]	*Krieger* et al. [1971]	[1978]
Brown et al. [1957]	*Weitzman* et al. [1971]	*Terpstra* et al. [1978]
Eik-Nes and Clark [1958]	*Rose* et al. [1972]	*Yamaguchi* et al. [1978]
Perkoff et al. [1959]	*Dzieniszewski and*	*Lorenzo and Mancheño*
Nugent et al. [1960]	*Milewski* [1974]	[1979]
Halberg et al. [1961]	*Ghosh* et al. [1974a,b]	*Nathan* et al. [1979]
Krieger [1961]	*Grim* et al. [1974]	*Ziegler* et al. [1979]
Ney et al. [1963]	*Turton and Deegan* [1974]	*Guignard* et al. [1980]
Givens et al. [1964]	*Colucci* et al. [1975]	*Miyatake* et al. [1980]
Nichols et al. [1965]	*Cugini* et al. [1975]	*Morimoto* et al. [1980]
McClure [1966]	*Rosenfeld* et al. [1975]	*Nolten* et al. [1980]
Cade et al. [1967]	*Silber-Kasprzak* et al.	*Schell* et al. [1980]
Franks [1967]	[1975]	*Degli Uberti* et al. [1981]
Knapp et al. [1967]	*Tavadia* et al. [1975]	*Lightman* et al. [1981]
Orth et al. [1967]	*Van Cauter* et al. [1975]	*Sakamaki* et al. [1981]
Martin and Martin [1968]	*Ichikawa* et al. [1976]	*Walter-Van Cauter* et al.
Takahashi et al. [1968]	*Rastogi* et al. [1976]	[1981]
	Cugini et al. [1977]	

plasma 17-OHCS in man became obscure and the new rhythms synchronized with new sleep-wake patterns were observed. Mostly, plasma 17-OHCS levels were sharply elevated during sleeping and attained a peak around the time of awakening. In the above two studies the rooms were carefully darkened during sleeping. Thus, the effect of alteration in the sleep-wake pattern could not be separated from that of the light-dark pattern. To separate these two variables, *Krieger* et al. [1969] studied the circadian variation of plasma 11-OHCS levels in normal adults subjected for 3 weeks to normal daily activity and nocturnal sleep (sleep from midnight to 0800 hours) under constant light and for 13 days to a reversed sleep-wake pattern (sleep from 1200 to 2000 hours) under constant light. It was observed that normal daily activity and nocturnal sleep under constant light did not result in any change in circadian periodicity of plasma 11-OHCS, whereas a reversed sleep-wake pattern under constant light induced a phase reversal of circadian rhythm of plasma 11-OHCS levels. The above findings suggest that the sleep-wake rather than the light-dark pattern is a principal determinant in the periodicity in man of the adrenocortical secretory activity. In contrast, *Orth and Island* [1969] noted that the light-dark pattern might be important for producing the circadian periodicity of adrenocortical secretion.

In the study of the circadian periodicity of adrenocortical secretory activity the frequency of plasma sampling is an important factor. In normal subjects a relatively simple pattern of circadian variations of plasma corticosteroids characterized by a smooth rise and decline every 24-hour period is observed when plasma samplings are performed at relatively long intervals, such as 2–6 h. However, *Weitzman* et al. [1966] found a series of 3–4 sporadic elevations of plasma 17-OHCS in man during the latter half of sleeping period when they performed plasma sampling every 30 min during night. By sampling at 1-hour intervals, *Orth* et al. [1967] observed small irregular rises and declines of plasma 17-OHCS levels during a period of wakefulness. A similar finding was noted by *Orth and Island* [1969]. The sporadic variations of plasma cortisol levels during a 24-hour period have been extensively studies by *Hellman* et al. [1970]. In their study, determinations of specific activity of plasma cortisol following i.v. administration of [4-^{14}C] cortisol and those of plasma cortisol concentrations were performed at frequent intervals. It was found that the basal secretion of cortisol in man was an intermittent rather than a continuous process and each sporadic rise in plasma cortisol levels represented a brief episode of cortisol secretion which was followed by a period of no cortisol secretion. It was also observed that during a 24-hour period the cortisol secretion occurred only in 8 episodic periods which occupied about 6 out of 24 h. By sampling at 30-min intervals *Krieger* et al. [1971] showed a circadian pattern of plasma 11-OHCS levels with 5–10 peaks, most of which occurred in a 9-hour period (from midnight to 0900 hours). Episodic nature of cortisol secretion in man has also been shown by *Gallagher* et al. [1973], *Armbruster* et al. [1975], *Rosenfeld* et al. [1975] and *Walter-Van Cauter* et al. [1981].

B. Studies in Monkeys

Circadian periodicity of plasma cortisol (or corticosteroid) levels has been demonstrated in rhesus [*Migeon* et al., 1955; *Mason* et al.,

1957; *Michael* et al., 1974; *Krey* et al., 1975; *Setchell* et al., 1975b; *Ferin* et al., 1977; *Leshner* et al., 1978; *Rose* et al., 1978; *Spies* et al., 1979] and squirrel [*Wilson* et al., 1978] monkeys. The circadian plasma cortisol pattern in monkeys is similar to that in man; the peak levels were attained during early morning hours and the lowest levels in the evening.

In the above studies plasma samples were mostly obtained every 3–6 h. In rhesus monkeys, whose peripheral plasma samples were collected every 15 min, *Jacoby* et al. [1974] found that the appearance of episodic secretion of cortisol resembled that in man; secretory episodes appeared frequently during nocturnal and early morning hours.

C. Studies in Rats

The existence of the circadian periodicity of plasma corticosterone levels in rats has been thoroughly documented (table V). In normal rats fed ad libitum and kept under an ordinary lighting schedule, the lowest levels of plasma corticosterone mostly appear during early morning hours, progressive elevations during afternoon and the peak levels in the evening shortly before the onset of the dark period. It has been reported that an apparent circadian periodicity of plasma corticosterone in immature rats is first observed after birth at 30–32 days [*Allen and Kendall*, 1967], 21–25 days [*Ader*, 1969], 18 days [*Campbell and Ramaley*, 1974], 22 days [*Levin and Levine*, 1975], 24 days [*Ramaley*, 1975], 28 days [*Ramaley and Sieck*, 1975], 27 days [*Ulrich* et al., 1976], 18 days [*Lengvári* et al., 1977], around the end of the 4th week [*Takahashi* et al., 1979] and 29–30 days [*Kimura* et al., 1981].

Effect of Constant Lighting

Critchlow [1963] studied the circadian pattern of plasma and adrenal corticosterone levels in male and female rats exposed to constant light for 103 days. No apparent circadian variation with distinct peaks of corticosterone levels was observed in male rats, whereas a circadian periodicity with the earlier appearance of peaks was found in female rats. *Cheifetz* et al. [1968] showed that the circadian variation of plasma corticosterone levels in female rats was almost completely abolished when the animals were exposed to constant illumination for 8 weeks. *Krieger* [1973] noted that no normal circadian variation of plasma cor-

Table V. Reports of circadian variations of plasma corticosterone levels in the rat

Guillemin et al. [1959a]
McCarthy et al. [1960]
Slusher and Browning [1961]
Critchlow [1963]
Critchlow et al. [1963]
Saba et al. [1963b]
Slusher [1964]
Saba et al. [1965]
Tronchetti et al. [1965]
Holmquest et al. [1966]
Scheving and Pauly [1966]
Allen and Kendall [1967]
Halász et al. [1967]
Zimmermann and Critchlow [1967]
Ader and Friedman [1968]
Cheifetz et al. [1968]
Retiene et al. [1968]
Ader [1969]
David-Nelson and Brodish [1969]
Hiroshige et al. [1969a]
Palka et al. [1969]
Hiroshige and Sakakura [1971]
Moberg et al. [1971]
Scapagnini et al. [1971a]
Seggie and Brown [1971]
Takebe et al. [1972]
Hiroshige et al. [1973]
Hiroshige and Wada-Okada [1973]
Johnson and Levine [1973]
Krieger [1973]
Ulrich and Yuwiler [1973]
Wilson and Critchlow [1973/74]
Campbell and Ramaley [1974]

Jacobs [1974]
Krieger [1974]
Marotta et al. [1974]
Seggie et al. [1974]
Stern and Levine [1974]
Van Cantfort [1974]
Wiley et al. [1974]
Wilson and Critchlow [1974]
Fukuda and Greer [1975]
Fukuda et al. [1975]
Lanier et al. [1975]
Levin and Levine [1975]
Miyabo and Hisada [1975]
Moberg et al. [1975]
Morimoto et al. [1975]
Ramaley [1975]
Ramaley and Sieck [1975]
Simon and George [1975]
Butte et al. [1976]
Conlee et al. [1976]
Dunn et al. [1976a, b]
Gomez-Sanchez et al. [1976]
Hilfenhaus [1976]
Inoue et al. [1976]
Levin et al. [1976]
Takahashi et al. [1976]
Ulrich et al. [1976]
Wilson et al. [1976]
Yasuda et al. [1976]
Engeland et al. [1977]
Lengvári et al. [1977]
Lengvári and Liposits [1977a]
Mann et al. [1977]
Morimoto et al. [1977]
Obled et al. [1977]
Takahashi et al. [1977a, b]
Wilson and Greer [1977]

Armstrong and Hatton [1978]
Daiguji et al. [1978]
Dallman et al. [1978]
Dunn and Carrillo [1978]
Dunn and Johnson [1978]
Fatranská et al. [1978]
Guillemant et al. [1978]
Honma and Hiroshige [1978]
Sakakura et al. [1978]
Swan et al. [1978]
Young and Walker [1978]
Brownie et al. [1979]
Gallant and Brownie [1979]
Kobayashi and Takahashi [1979]
Morimoto et al. [1979]
Rookh et al. [1979]
Szafarczyk et al. [1979]
Takahashi et al. [1979]
Wilkinson et al. [1979]
Ahlers et al. [1980]
Allen-Rowlands et al. [1980]
Guillemant et al. [1980]
Hilfenhaus and Herting [1980]
Holbrook et al. [1980]
Kaneko et al. [1980]
Kato et al. [1980]
Lengvári and Szelier [1980]
Seggie [1980]
Szafarczyk et al. [1980a, c]
Gibson and Krieger [1981]
Itoh et al. [1981]
Kimura et al. [1981]
Miyabo et al. [1981]
Poland et al. [1981]
Szafarczyk et al. [1981]
Vernikos et al. [1982]

ticosterone levels was observed in male and female rats after exposure to constant illumination for several weeks. *Morimoto* et al. [1975] and *Takahashi* et al. [1977b] observed in female rats a phase shift of circadian rhythm of plasma corticosterone (or 11-OHCS) after 2 weeks of exposure to constant light and an abolition of the circadian periodicity after 3 or 4–7 weeks. *Wilson and Greer* [1977] reported that non-synchronized circadian variation in plasma corticosterone levels was observed in 5 out of 9 female rats kept under the constant light conditions for 15 weeks and arrhythmic variation in the other 4 rats. *Honma and Hiroshige* [1978] noted a marked phase shift in circadian periodicity of plasma corticosterone levels in male rats after 12 days of constant light exposure. A phase-shifted, inverted pattern of circadian periodicity in plasma corticosterone in female rats after 10-day exposure to constant light was observed also by *Morimoto* et al. [1979].

The effect of constant light on the circadian rhythm of plasma corticosterone in prepubertal rats has been extensively studied by *Ramaley* [1975]. Male and female rats born and maintained under constant light did not show at 24 or 28 days of age any distinct circadian periodicity in plasma corticosterone levels, whereas a regular circadian rhythm with peak corticosterone levels at 1800 hours was observed in 24- or 28-day-old rats born and raised under the standard light cycle (lights on at 0500 and off at 1900 hours). In rats raised in constant light and then placed at weaning (21 days of age) in the standard light cycle, no circadian rhythm in plasma corticosterone levels was observed at 24 days; however, a significant periodicity with a peak at 1800 hours was found at 28 days. In rats raised in the standard light cycle and placed in constant light at weaning no circadian rhythm of plasma corticosterone appeared at 24 or 28 days of age.

Effect of Constant Darkness

In male rats maintained under conditions of constant darkness for 3 weeks, daily variations of plasma corticosterone levels in individual animals show a free-running pattern [*Scheving and Pauly,* 1966]. No normal circadian periodicity is observed in male and female rats maintained under constant dark conditions for 50 days [*Krieger,* 1973]. However, *Morimoto* et al. [1975, 1977, 1979] failed to find any significant effect of constant darkness (for 7–21 days or 10 days) on circadian rhythm of plasma 11-OHCS (or corticosterone) levels in female rats.

Effect of Blindness

The circadian rhythm of plasma corticosterone levels has been found to be abolished in blinded male rats 7 days [*Saba* et al., 1965; *Tronchetti* et al., 1965] and 1–6 weeks [*Jacobs,* 1974] after ocular enucleation. A phase-shifted circadian periodicity of plasma corticosterone in male rats 30 days after enucleation [*Saba* et al., 1965] has also been reported. Phase-shifted patterns of circadian variation of plasma corticosterone levels after blinding have been extensively studied in male [*Takahashi* et al., 1976] and female [*Takahashi* et al., 1977b] rats. In male rats a 4-hour delay of peak elevation of plasma corticosterone levels was observed at the end of the 4th week and a phase reversal of circadian rhythm of plasma corticosterone between the 8th and 10th week after enucleation. In female rats a phase shift appeared after the 3rd week and the circadian rhythm with an inverted phase at the end of the 7th week. Similar observations were noted by *Szafarczyk* et al. [1980c]. *Krieger* [1973] reported that in male and female rats blinded at 1 day of age a phase shift with a lower amplitude of circadian variations of plasma corticosterone was observed at 30 days of age and aperiodic peaking with elevated mean plasma levels of corticosterone at 80 days. In 80-day-old rats enucleated at 30 days of age the pattern with aperiodic peaking and normal mean plasma levels was observed. In blinded female rats *Wilson* et al. [1976] found at 3 weeks after enucleation a synchronized circadian variation of plasma corticosterone comparable to that in intact rats and at 10 weeks arrhythmic fluctuations of mean plasma corticosterone levels. At 10 weeks, however, the fluctuations of plasma corticosterone of each rat were synchronized when the peak values on the 1st day of observation in individual rats were aligned at 0 time. These findings suggest that individual blinded rats possess free-running circadian rhythms which are synchronized neither with those of other blinded rats nor with the period of transition from light to dark. *Gibbs* [1976] has demonstrated that free-running circadian rhythms of plasma corticosterone levels in individual blinded female rats have the same phase relationship to the activity cycle as in intact rats; the peak of plasma corticosterone levels appears at the beginning of the running phase. Free-running circadian variations of plasma or blood corticosterone have been observed in 45-day-old female rats blinded at 1 day of age, in 84-, 112- and 142-day-old rats blinded at 1, 26 or 60 days of age [*Swan* et al., 1978] and 4- to 6-week-old rats blinded within 24 h after birth [*Takahashi* et al., 1979].

Effect of Altered Light Cycle

In male and female rats, which were kept under normal light-dark alterations (lights on at 0400, off at 1800 hours) for 3 weeks and then subjected to a 9-hour shift in the lighting period (lights on at 1300, off at 0300 hours) for 3 weeks *Critchlow* et al. [1963] have observed a phase shift of about 9 h of plasma and adrenal corticosterone rhythms. In male rats a 12-hour shift of light and dark periods (inverted lighting schedule) for 3 weeks or more seems to reverse the circadian rhythm of plasma corticosterone [*Scheving and Pauly*, 1966]. Similar results have been obtained in male rats by *Marotta* et al. [1974]. A complete reversal of the circadian rhythm of plasma corticosterone (or 11-OHCS) is induced in female rats by reversing 12-hour periods of light and darkness for 14–15 days [*Van Cantfort*, 1974; *Morimoto* et al., 1975] and only for 3 days [*Morimoto* et al., 1977].

In male rats *Holmquest* et al. [1966] gave light and darkness in a randomized pattern but in the same length of total light and total dark periods per 24 h for 17–40 days. Synchronized circadian periodicities in the mean plasma and adrenal corticosterone levels were attenuated and the circadian rhythms in individual rats showed free-running patterns.

Effect of Restriction of Feeding (or Drinking) Time

In female rats maintained under a standard light-dark cycle (lights on at 0700 and off at 1900 hours), *Johnson and Levine* [1973] have shown that when the watering period is restricted to 1 h for 60–102 days a significant elevation of plasma corticosterone appears 1 h before the onset of drinking time and the evening peak of plasma corticosterone is markedly suppressed. *Krieger* [1974] reported that in rats maintained under normal lighting conditions (lights on at 0800 and off at 2000 hours) a 12-hour shift in the time of the peak of plasma corticosterone levels and a marked increase in daytime running activity were observed when access to food and water was restricted to a 2-hour period (0930 to 1130 hours) for 15 days. The finding suggests that an alteration of the sleep-wake pattern induced by restriction of feeding and drinking time might be a factor for the change in circadian rhythm of plasma corticosterone.

In male rats maintained under normal lighting conditions (lights on at 0500 and off at 1900 hours), *Moberg* et al. [1975] found 2 circadian peaks of plasma corticosterone levels after the feeding time of the animals was restricted to a 2-hour period (from 0800 to 1000 hours) for 2 weeks; one peak was observed just prior to feeding time and a second small peak just before the time of lights off. Similar findings were reported also by *Inoue* et al. [1976]. In male rats maintained under a normal light-dark cycle (lights on at 0700 and off at 1900 hours) feeding time was restricted to an 8-hour period (1000–1800 hours) for 2 weeks. A phase-shifted circadian rhythm with 2 peaks of plasma corticosterone levels, one peak at 0800 hours and a second small peak at 1800 hours, was observed.

The effect of feeding time restriction on circadian variations of plasma 11-OHCS in the rat has been extensively studied by *Morimoto* et al. [1977]. In female rats maintained under normal light-dark conditions (lights on at 0800 and off at 2000 hours), feeding time was restricted to a 6-hour period in 4 different schedules; food and water were given only during the first 6 h (group A) or the last 6 h (group B) of the 12-hour dark period, or only during the first 6 h (group C) or the last 6 h (group D) of the 12-hour light period. A single circadian peak of plasma 11-OHCS levels was found just or 3 h before feeding time in groups A and B, respectively. Two peaks, however, appeared in groups C and D. In group C one peak was observed just prior to feeding time and the other just prior to the dark period. In group D one peak appeared at 3 h prior to feeding time and a second small peak at 3 h prior to the onset of the dark period. They stated that the periodicity in rest-activity, sleep-waking or fasting-eating might be the more important and potent synchronizer (Zeitgeber) of the circadian periodicity of the adrenocortical secretory activity than the light-dark cycle.

Takahashi et al. [1977a] reported that in female rats maintained under continuous illumination for 5 weeks, the circadian periodicity of the plasma corticosterone levels was abolished when the animals were allowed to feed ad libitum. In these rats, however, when access to food was restricted to a 2-hour period (0900–1100 hours), the circadian peak of plasma corticosterone levels just prior to the feeding time reappeared within 1 week. In male rats maintained under normal light-dark conditions (lights on at 0900, off at 2100 hours), *Wilkinson* et al. [1979] showed that the peak plasma corticosterone levels appeared just prior to the feeding time when access to food was restricted to a 2-hour period. Similar findings were reported by *Ahlers* et al. [1980], *Itoh* et al. [1980] and *Kato* et al. [1980].

Inoue et al. [1976] and *Kobayashi and Takahashi* [1979] reported that normal circadian periodicity of plasma corticosterone could be observed in male rats during 3 days of total food deprivation (water was given ad libitum); however, it was found to be abolished after a 5-day deprivation.

D. Studies in Other Animal Species

The occurrence of circadian variation of plasma corticosterone in mice similar to that in rats has been reported [*Halberg* et al., 1959a,b; *Galicich* et al., 1963; *Haus* et al., 1967; *Smolensky* et al., 1978]. Circadian periodicity of plasma cortisol in guinea pigs [*Garris*, 1979] and in rabbits [*Singh* et al., 1975, 1977] has also been noted.

Harwood and Mason [1956] found that the peripheral plasma levels of 17-OHCS in the dog at 0900 hours were higher than those at 1600 hours. *Rijnberk* et al. [1968] observed a circadian variation of peripheral plasma levels of 11-OHCS in dogs with the highest values around 0800 and the lowest around 2300 hours. *Arcangeli* et al. [1973] showed a typical circadian variation of peripheral plasma corticosteroid levels, the peaks of which were attained at morning hours. By contrast, *Kuipers* et al. [1958] and *Johnston and Mather* [1978] failed to find a circadian periodicity of peripheral plasma 17-OHCS or cortisol levels in dogs. In cats, *Krieger and Krieger* [1967] and *Krieger and Rizzo* [1969] have shown that the peak levels of peripheral plasma 17-OHCS are attained at some time between midnight and 0400 hours.

The circadian periodicity and episodic nature of changes in peripheral plasma cortisol levels in sheep [*McNatty* et al., 1972; *McNatty and Thurley*, 1973; *Fulkerson and Tang*, 1979] and the circadian variation of peripheral plasma cortisol levels in the horse [*James* et al., 1970; *Larsson* et al., 1979] have also been reported.

Chapter 6. Participation of the Hypothalamus in the Control of Pituitary-Adrenocortical Activity

A. Hypothalamic Stimulation

Endrőczi et al. [1956] have shown that the adrenal ascorbic acid concentration in rats is markedly reduced after electrical stimulation of the tuber cinereum and mammillary body, and only slightly decreased by stimulation of the ventral hypothalamic nuclei.

Suzuki et al. [1960] studied the effect of electrical stimulation of the diencephalon on adrenal 17-OHCS secretion in conscious dogs. A marked increase in adrenal 17-OHCS secretion was found to be induced by stimulation of the ventral posterior hypothalamic areas, whereas a slight initial increase followed by a decrease was observed after stimulation of the dorsal posterior hypothalamic areas. *Katsuki* [1961] evaluated adrenal 17-OHCS secretion in dogs in response to electrical stimulation of various sites of the hypothalamus. In his experiments 3 different patterns of responses in adrenocortical secretion were observed after stimulation of the anterior, mid- and posterior hypothalamic areas. An increase in adrenal 17-OHCS secretion in pentobarbital-anesthetized dogs induced by electrical stimulation of the ventral hypothalamic areas [*Goldfien and Ganong*, 1962; *Ganong* et al., 1965] and of various regions of the hypothalamus [*Ishihara* et al., 1965a] has also been noted.

In hexobarbital-anesthetized cats *Katsuki* et al. [1955] have observed a marked increase in chemocorticoid content of adrenal venous blood samples after electrical stimulation of the posterior hypothalamic nucleus, mammillary body and their adjacent areas. *Katsuki* [1961] also showed an increase in adrenal secretion of 17-ketogenic steroids in pentobarbital-anesthetized cats induced by electrical stimulation of the mid- and posterior hypothalamus. In pentobarbital-anesthetized cats, *Endröczi and Lissák* [1963] observed that a marked increase in adrenal corticosteroid secretion rate was elicited by electrical stimulation of the posterior tuber cinereum, median eminence and pre-mammillary region, whereas a moderate decrease was induced by the stimulation of the lateral hypothalamus.

Electrical stimulation of the hypothalamus has been found to elevate peripheral plasma corticosteroid levels in rhesus monkeys [*Mason, 1958; Natelson* et al., 1974], in rats [*Dunn and Critchlow, 1973; Redgate and Fahringer, 1973; Makara and Stark, 1976*], in dogs [*Salcman* et al., 1970] and in cats [*Kokka* et al., 1972].

Preoptic Stimulation

Suzuki et al. [1960] have observed in 1 conscious dog a slight increase followed by a delayed decrease in adrenal 17-OHCS secretion rate after electrical stimulation of the preoptic area. In 1 acute encéphale isolé cat preparation, *Slusher and Hyde* [1961b] have demonstrated a marked decrease in adrenal effluent corticosteroid levels after stimulation of the preoptic region. *Endröczi and Lissák* [1963] showed in 2 pentobarbital-anesthetized cats that electrical stimulation of the lateral preoptic area markedly depressed the adrenal corticosteroid secretion. In 5 pentobarbital-anesthetized cats *Yoshio* [1964] observed that the adrenal 17-OHCS secretion rates were reduced by 24–32% after electrical stimulation of the medial preoptic area. In bilaterally vagotomized encéphale isolé cat preparations *Taylor and Branch* [1969] found that electrical stimulation of the preoptic-basal forebrain region decreased the adrenocortical secretion rate by $37.6 \pm 7.3\%$. In contrast to the above findings, *McHugh* et al. [1966] noted that peripheral plasma levels of 17-OHCS in conscious monkeys were markedly elevated by electrical self- or automatic stimulation of the lateral preoptic area. *Redgate and Fahringer* [1973] also found a small but significant elevation of plasma corticosterone levels in rats induced by electrical stimulation of the preoptic area.

B. Pituitary Stalk Section

Studies of the effect of the pituitary stalk section on the pituitary-adrenocortical secretory response to stressful stimuli have revealed disparate results; the discrepancies might at least in part be due to species difference.

Adrenal ascorbic acid depletion in rats 1 h after exposure to surgical stress (unilateral adrenalectomy under ether anesthesia) has been found to be abolished by complete section [*McCann*, 1953] and lesions [*Kovács* et al., 1962] of the pituitary stalk. The latter investigators have also observed that adrenal corticosterone secretion rates under conditions of surgical stress in stalk-sectioned rats are significantly lower than those in intact rats. *Matsuda* et al. [1963] reported that adrenal corticosterone secretion in rats in response to stressful stimuli (a leg break under ether anesthesia) was markedly depressed by pituitary stalk section.

By contrast, adrenal cholesterol depletion following exposure for 18 h to cold (5 °C) in stalk-sectioned guinea pigs is not significantly different from that in intact animals [*Tang and Patton,* 1951]. *Fortier* et al. [1957] showed that the adrenal ascorbic acid concentration in rabbits – whose pituitary stalk was sectioned, and little, if any, regeneration of the portal vessels was observed – was depleted 1.5 h after exposure to surgical stress (unilateral adrenalectomy under ether anesthesia) from the mean initial value of 269 ± 20 mg/100 g of adrenal weight to 215 ± 15 mg/100 g, a significant depletion comparable to that in intact control animals.

Adrenal 17-OHCS secretion in ether-anesthetized dogs in response to electrical stimulation of the sciatic nerve and that in response to laparotomy are not blocked by pituitary stalk section [*Hume and Egdahl,* 1959]. *Zukoski and Ney* [1966] showed that in pentobarbital-anesthetized stalk-sectioned dogs the adrenal 17-OHCS secretion rates increased after laparotomy from the baseline levels of 1.6 ± 0.2 to 8.6 ± 2.1 µg/min and this adrenocortical secretory response was comparable with that in intact dogs. They also noted that adrenal 17-OHCS secretion in response to sensory nerve stimulation (traction of the sciatic nerve) in pentobarbital-anesthetized stalk-sectioned dogs seemed smaller than, but was not significantly different from, that in control dogs. *Wise* et al. [1963] reported that adrenal 17-OHCS secretion rates following exposure to surgical stress (adrenal vein cannulation) in pentobarbital-anesthetized dogs 4 h after pituitary stalk section were 5.9 ± 1.6 µg/min, whereas those in control animals were 10.8 ± 2.0 µg/min. *Kendall and Roth* [1969] observed that peripheral plasma 11-OHCS levels in chronically stalk-sectioned monkeys were not below the normal range, and in 1 of these animals plasma 11-OHCS levels were promptly elevated following exposure to surgical stress (oophorectomy).

C. Hypothalamic Lesions

Adrenal ascorbic acid depletion in rats following exposure to surgical stress (unilateral adrenalectomy under ether anesthesia) has been
found to be prevented by lesions of the median eminence [*McCann,*
1953; *McCann and Haberland,* 1960]. *McCann and Brobeck* [1954] have
observed that destruction of the supraopticohypophyseal tract in the
rat, as evidenced by the location of lesions and the presence of diabetes
insipidus, decreases or abolishes the adrenal ascorbic acid depletion in
response to surgical trauma (unilateral adrenalectomy under ether
anesthesia) with or without i.p. administration of histamine or epinephrine. *Endröczi and Mess* [1955] and *Slusher and Roberts* [1956]
have noted that the hypothalamic lesions involving the tuber cinereum
and mammillary bodies (or anterior border of the mammillary bodies)
completely block the adrenal ascorbic acid depletion in response to
surgical stress. Lesions of the median eminence in rats have been
shown to abolish the adrenal ascorbic acid depletion induced by i.p.
administration of aspirin [*George and Way,* 1957; *Way* et al., 1962],
lipid extract of bovine posterior hypothalamus [*De Wied* et al., 1958]
and morphine [*George and Way,* 1959]. *Slusher* [1958] has demonstrated that the adrenal ascorbic acid depletion in the rat in response to
surgical stress (laparotomy followed by adrenal vein cannulation) is
inhibited by lesions in the basal tuberal region but not by lesions in the
posterior and mid-central portions of the hypothalamus. *Brodish* [1964]
showed that the adrenal ascorbic acid depletion in rats 1 h but not 2 h
after surgical trauma (unilateral adrenalectomy) was abolished by
placement of lesions in any of the anterior, middle and posterior regions of the ventral hypothalamus. It was also shown that larger hypothalamic lesions produced a longer delay, but not a complete block, of
the adrenal ascorbic acid response to surgical stress.

Direct evaluation of adrenocortical secretion in rats with hypothalamic lesions in response to stress has been performed by *Slusher* [1958].
She has observed that the adrenal corticosterone secretion in response
to surgical stress (laparotomy plus left adrenal vein cannulation) is inhibited by lesions involving the posterior and mid-central portions of
the hypothalamus but not by lesions in the basal tuberal regions. *Porter*
[1963] determined the secretion rates of corticosterone in rats with lesions in the anterior hypothalamus, the centers of lesions being equi-
distant from the posterior margin of the optic chiasm and the pituitary

stalk, and in intact rats after adrenal vein cannulation under chloralose-urethane anesthesia. In intact rats the rate of corticosterone secretion was 10.7 ± 2.3 µg/15 min for the first 15-min interval, which increased steadily and attained 17.8 ± 0.6 µg/15 min at the 7th 15-min interval, whereas in rats with hypothalamic lesions the secretory rate of
corticosterone was 1.8 ± 0.4 and 11.1 ± 1.7 µg/15 min, respectively.
Thus, it is suggested that the adrenocortical secretory response in rats
to surgical stress is suppressed by hypothalamic lesions. *Porter* [1969]
studied the effect of lesions in the SME (the junction of the stalk and
median eminence) and in the AME (the anterior median eminence) on
the adrenal corticosterone secretion response to surgical stress (adrenal
vein cannulation under chloralose-urethane anesthesia). Hypothalamic lesions in the SME completely suppressed the corticosterone secretion response to surgical stress throughout a 70-min period of observation, whereas hypothalamic lesions in the AME only slowed the rate
at which the secretion of corticosterone was increasing to the maximum levels.

Hume and Nelson [1955b] have evaluated the effect of hypothalamic lesions on adrenal 17-OHCS secretion following surgical stress
(adrenal vein cannulation) in ether-anesthetized dogs. The rate of
17-OHCS secretion in dogs with lesions in the AME averaged 1.8
µg/min, whereas the mean secretion rate in intact dogs was 12.5
µg/min. *Daily and Ganong* [1958] have studied the adrenal 17-OHCS
secretion in dogs with hypothalamic lesions in response to surgical
stress (adrenal vein cannulation under pentothal-ether anesthesia). In 8
of 18 dogs with ventral hypothalamic lesions the adrenal 17-OHCS secretion rate was less than 3.75 µg/min, whereas in 15 intact dogs the secretion rate averaged 6.9 ± 0.5 µg/min. In 8 dogs with depressed
17-OHCS secretion response, various portions of the area between the
optic chiasm and the anterior edge of the mammillary bodies were
found to be destroyed. An area common to lesions in all these dogs was
the anterior end of the median eminence. *Ganong* et al. [1959] reported
the data of their experiments in which the adrenal 17-OHCS secretion
in response to surgical stress was measured in 4 groups of dogs after
acute adrenal vein cannulation under pentothal-ether anesthesia. In
dogs with an intact hypothalamus the adrenal 17-OHCS secretion rate
was 6.5 ± 0.8 µg/min, whereas in dogs with lesions in the AME the secretion rate was 2.9 ± 0.3 µg/min. It was 7.3 ± 0.9 µg/min in dogs with
lesions in other portions of the hypothalamus. Thus, it was demon-

strated that the adrenocortical secretion in the dog in response to surgical stress was depressed by lesions in the AME but not by lesions in other hypothalamic areas. *Hume and Egdahl* [1959] showed that the adrenal 17-OHCS secretion in dogs in response to electrical stimulation of the sciatic nerve was abolished by removal of the anterior hypothalamic and preoptic areas. *Ganong* et al. [1961] found that the adrenal secretion of cortisol and corticosterone in ether-anesthetized dogs was suppressed by lesions in the median eminence but it was not affected by lesions of other portions of the hypothalamus.

In pentobarbital-anesthetized dogs the effect of anterior and posterior hypothalamic lesions on adrenocortical secretory responses to surgical stress, histamine and metyrapone (SU-4885) has been studied by *Katsuki* et al. [1967]. In order to evaluate the adrenocortical secretory response to surgical stress, one adrenal venous blood sample was collected immediately after the completion of adrenal vein cannulation. After 20–24 h, when the animals recovered from surgical trauma, adrenal 17-OHCS secretion rates before and after i.v. injections of histamine and metyrapone were determined; the latter drug has been known to induce a depression of circulating cortisol levels and thus stimulate the adrenal secretion of 17-OHCS via the pituitary ACTH release. In dogs with lesions in the anterior hypothalamus involving the supraoptic, paraventricular, ventromedial and dorsomedial nuclei of the hypothalamus, the adrenal 17-OHCS secretion in response to surgical stress, histamine and metyrapone was suppressed. In dogs with lesions in the posterior hypothalamus involving the lateral mammillary and premammillary nuclei and the dorsal portion of the supramammillary nuclei, the adrenocortical secretory responses to surgical stress and histamine were not significantly different from those in normal controls, whereas the response to metyrapone was suppressed. The experimental results suggest that the posterior hypothalamus participates in a negative feedback mechanism for regulation of the pituitary ACTH release, while the anterior hypothalamus is involved in the central nervous system mechanism of the pituitary-adrenocortical activation in response to stressful stimuli.

Suzuki et al. [1975a] examined the effect of hypothalamic lesions on adrenocortical secretory response in dogs to pilocarpine. The experimental results are shown in figure 26. In intact pentobarbital-anesthetized dogs, in which adrenal vein cannulation was performed approximately 18 h before the start of observation, the adrenal 17-OHCS

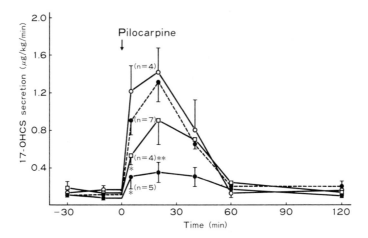

Fig. 26. Effect of hypothalamic lesions on pilocarpine-induced adrenocortical secretion in pentobarbital-anesthetized dogs. Adrenal 17-OHCS secretion rates in intact
dogs (○) and in dogs with hypothalamic lesions in the anterior median eminence
(●——●), anterior ventromedial hypothalamic areas adjacent to the anterior median
eminence (□) and in other hypothalamic areas (●- - -●) before and after i.v. injection of
0.3 mg/kg pilocarpine hydrochloride. Vertical lines indicate the SEM. * p < 0.05, ** p <
0.01 vs intact dog group (Student's t-test). [Based on data from *Suzuki* et al., 1975a.]

secretion rate increased markedly after i.v. injection of pilocarpine (0.3
mg/kg). The adrenocortical secretory response in the dog to pilocarpine was found to be completely abolished by hypophysectomy. In
dogs with lesions of the AME, which extends from the posterior border
of the optic chiasm to the anterior border of the SME, the adrenal cortical response to pilocarpine was significantly impaired; the secretion
rates at 5 and 20 min after the injection of pilocarpine in these dogs
with AME lesions were significantly smaller than the corresponding
values in intact dogs. In dogs with lesions of the anterior ventromedial
hypothalamic regions adjacent to the AME a significant suppression
of adrenocortical secretory response to pilocarpine was observed only
at 5 min after the injection. At 20 min after pilocarpine injection the secretion rates in these dogs seemed smaller than, but were not significantly different from, those in intact dogs. In dogs with lesions of other
hypothalamic regions, involving the anterior ventrolateral hypothalamic areas, the mammillary bodies and a part of the supraoptic areas,

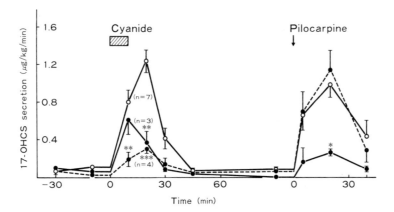

Fig. 27. Effect of hypothalamic lesions on adrenocortical secretory responses to cyanide and pilocarpine in pentobarbital-anesthetized dogs. Adrenal 17-OHCS secretion rates in intact dogs (O), in dogs with lesions in the anterior median eminence (●——●) and posterior median eminence (●- - -●) before and after a 10-min i.v. infusion of 1.0 mg/kg potassium cyanide and i.v. injection of 0.3 mg/kg pilocarpine hydrochloride. Vertical lines indicate the SEM. * $p < 0.05$, ** $p < 0.01$, *** $p < 0.001$ vs intact dogs group (Student's t-test). [Based on data from *Suzuki* et al., 1975b.]

the adrenal 17-OHCS secretion in response to pilocarpine was not significantly different from that in intact dogs.

Suzuki et al. [1975b] evaluated the effect of lesions of various areas of the hypothalamus on the adrenal 17-OHCS secretion in the dog in response to cyanide and pilocarpine. Data are illustrated in figures 27 and 28. In intact dogs a marked increase in adrenal 17-OHCS secretion was observed after i.v. infusion of cyanide (1.0 mg/kg) over 10 min as well as after i.v. injection of pilocarpine (0.3 mg/kg). The adrenocortical secretory responses to cyanide and pilocarpine were found to be completely blocked by hypophysectomy. In dogs with AME lesions a partial depression of the adrenocortical secretory response to cyanide and a marked impairment of the response to pilocarpine were observed. In dogs with lesions of the PME (the posterior median eminence), which extends from the anterior border of the SME to the anterior border of the premammillary areas, the secretory response to cyanide was markedly impaired, whereas the adrenal 17-OHCS secretion in re-

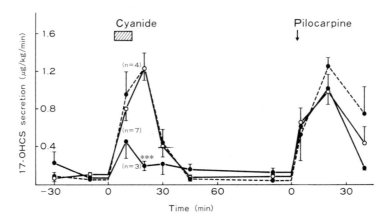

Fig. 28. Effect of hypothalamic lesions on adrenocortical secretory responses to cyanide and pilocarpine in pentobarbital-anesthetized dogs. Adrenal 17-OHCS secretion rates in intact dogs (○) and in dogs with lesions in the areas just dorsal to the posterior median eminence (●——●) and in the mammillary bodies (●- - -●) before and after a 10-min i.v. infusion of 1.0 mg/kg potassium cyanide and i.v. injection of 0.3 mg/kg pilocarpine hydrochloride. Vertical lines indicate the SEM. *** $p < 0.001$ vs intact dog group (Student's t-test). [Based on data from *Suzuki* et al., 1975b.]

sponse to pilocarpine was almost the same as that in intact dogs. In dogs with lesions of the areas just dorsal to the PME, the adrenocortical secretory responses to cyanide and pilocarpine were similar with those in dogs with lesions in the PME. Thus, in dogs with median eminence lesions a dissociation of adrenocortical secretory responses to cyanide and pilocarpine was demonstrated. In dogs with mammillary body lesions, no significant impairment of adrenocortical secretory responses to cyanide and pilocarpine was observed. The above findings suggest that the final pathway to the anterior pituitary for the adrenocortical secretory response to pilocarpine may traverse mainly the AME, whereas that for the secretory response to cyanide mainly the PME and partly the AME. Subsequently, *Hirose* et al. [1976] studied the effect of lesions of various hypothalamic areas on adrenal 17-OHCS secretion in the dog in response to histamine and cyanide. Adrenal 17-OHCS secretion was markedly increased by histamine (0.1 mg/kg, i.v. injection) and cyanide (1.0 mg/kg, i.v. infusion over 10 min). The adrenocortical secretory response to cyanide was completely abolished by hypophysectomy, whereas a small yet significant increase in adrenal 17-OHCS secretion was observed in hypophysectomized dogs after the injection of histamine. The latter finding suggests a possible direct action of histamine on the adrenal cortex. The adrenocortical secretory response to histamine was partially reduced by AME and PME lesions but it was not significantly suppressed by supramammillary lesions. However, the adrenocortical secretory response to cyanide was markedly suppressed by PME lesions. It was partially impaired by AME lesions and by supramammillary lesions. In dogs with premammillary lesions the adrenocortical secretory responses to cyanide and histamine were not different from those in intact dogs. The above findings, i.e. a partial dissociation of adrenocortical secretory responses in dogs with hypothalamic lesions to histamine and cyanide, suggest that the mechanism and the pathway to the anterior pituitary for the adrenocortical secretory response to histamine are different from those for the secretory response to cyanide.

To examine the effect of hypothalamic lesions on stress-induced activation of the pituitary-adrenocortical system, a number of investigators have used the peripheral plasma corticosteroid levels as an index for evaluating the adrenocortical secretory activity.

Brodish [1963] has used plasma corticosterone response in the rat to ether stress to determine the extent of the hypothalamic region which participates in the regulation of the pituitary-adrenocortical secretory activity. The ventral hypothalamic area, extending from the optic chiasm to the mammillary bodies, was divided into 4 zones and the effect of a single lesion in any of the zones and that of larger lesions extending over 2, 3 or 4 zones on the plasma corticosterone response to ether stress were studied. A single zone lesion was found to significantly impair the response; placement of larger lesions resulted in greater impairment of the response. The findings suggest that the whole area, rather than a localized region, of the ventral hypothalamus is involved in the central nervous system mechanism for the regulation of the pituitary-adrenocortical secretory activity. Subsequently, Brodish [1964] studied the effects of small and large hypothalamic lesions in rats on plasma corticosterone response to surgical stress (unilateral adrenalectomy). The response to surgical stress was markedly impaired in rats with small ventral hypothalamic lesions at 1 h but not 2 h after unilateral adrenalectomy. In rats with larger ventral hypothalamic lesions, the plasma corticosterone response was impaired for 2 h but not for 4 h following unilateral adrenalectomy. The experimental results indicate that the ventral hypothalamic lesions prevent the rapid response, but do not block the delayed response of the pituitary-adrenocortical system to surgical stress. Further studies on the delayed plasma corticosterone response in rats to stress, which was not blocked by ventral hypothalamic lesions, were done by *Brodish* [1969]. In rats with large ventral hypothalamic lesions extending over the posterior three quarters of the area from the optic chiasm to the mammillary bodies, plasma levels of corticosterone remained low during the first 2-hour period after the onset of exposure to cold (2 °C) or surgical stress (laparotomy under ether anesthesia), whereas a significant elevation of plasma corticosterone levels was observed at 4 and 6 h after the onset of stress. Intact rats could, but lesioned rats could not, sustain the elevated levels of plasma corticosterone for 24 h after the onset of stress. Application of a combination of two different stimuli, such as ether stress and exposure to cold, shortened the time for the onset of plasma corticosterone elevation from 4 to 2 h.

Plasma corticosterone response in rats to sound stress or 1-min electrical stimulation of the posterior diencephalon is inhibited by lesions of the posterior hypothalamus involving the posterior tuber cinereum but not by anterior hypothalamic lesions [*Slusher,* 1964]. In intact rats an i.p. injection of typhoid-paratyphoid vaccine results in a marked increase in plasma concentrations of corticosterone and this plasma corticosterone response is completely blocked by lesions of the median eminence [*Moberg,* 1971]. A rise of plasma corticosterone levels in rats after exposure to ether stress is abolished by lesions of the ventral hypothalamus extending from the retrochiasmatic area to the anterior border of the mammillary bodies [*Witorsch and Brodish,* 1972]. Plasma corticosterone response in rats to surgical stress (unilateral sham adrenalectomy under ether anesthesia) or a small dose of formaldehyde (2.5 mg/100 g, s.c.), but not to a large dose of formaldehyde (25 mg/100 g, s.c.) or *E. coli* endotoxin (25 μg/100 g, i.p.), has been found to be abolished after surgical ablation of the medial hypothalamus extending from the optic chiasm to the mammillary bodies [*Stark* et al., 1973/74].

Peripheral plasma 17-OHCS levels in intact dogs are significantly raised by psychic stress of a 2-hour immobilization – a response completely abolished by lesions of the hypothalamus involving the median eminence [*Ganong* et al., 1955b]. In squirrel monkeys, an elevation of plasma cortisol levels in response to capture and immobilization (chair restraint) has been observed to be blocked by posterior median eminence lesions [*Brown* et al., 1971].

The effect of hypothalamic lesions on the circadian periodicity of adrenocortical secretory activity in rats has been studied by some investigators. The circadian periodicity of plasma corticosterone levels in rats has been found to be impaired or abolished by lesions of the median eminence [*Saba* et al., 1963a; *Tronchetti* et al., 1965], the anterior hypothalamus involving the periventricular zone and arcuate nuclei but not the posterior tuber cinereum [*Slusher,* 1964], the ventromedial or dorsomedial hypothalamus [*Bellinger* et al., 1976] and the suprachiasmatic nuclei [*Krieger* et al., 1977a; *Raisman and Brown-Grant,* 1977; *Abe* et al., 1979]. *Moore and Eichler* [1972] noted that the circadian variation of the adrenal corticosterone in rats was abolished by suprachiasmatic lesions. *Krieger* et al. [1977a] reported that the circadian rhythmicity of plasma corticosterone was absent in rats with suprachiasmatic lesions when the animals were maintained under unrestricted access to food and water, whereas it was restored in these rats when they were kept under conditions of restricted feeding.

D. Hypothalamic Deafferentation

Hypothalamic deafferentation technique provided by *Halász and Pupp* [1965] has made it possible to produce transections of neural connections to the medial basal hypothalamus in the rat leaving the median eminence in intact anatomical connection with the pituitary gland and without removing any brain tissue. Using this technique they have shown that the adrenal ascorbic acid depletion in rats in response to surgical stress (unilateral adrenalectomy under hexobarbital anesthesia) is mostly not impaired by complete deafferentation of the medial basal hypothalamus. *Halász* et al. [1967] have reported that the plasma corticosterone levels are elevated following surgical stress of unilateral adrenalectomy under ether anesthesia in intact rats; in rats with complete deafferentation of the medial basal hypothalamus, however, changes of plasma corticosterone levels in response to surgical stress are not significant. Subsequently, a number of studies have been done in hypothalamic-deafferented rats by some groups of investigators to determine the extent of the neural connections to the medial basal hypothalamus which participate in the stress-induced pituitary-adrenocortical activation. Experimental results obtained in rats with complete deafferentation of the medial basal hypothalamus are summarized in table VI and those in rats with partial hypothalamic deafferentation in table VII.

The findings obtained in experiments on rats with complete hypothalamic deafferentation suggest that the stimulatory effects on the pituitary adrenocortical system in rats of audiogenic or olfactory stimulation, sciatic nerve stimulation, leg break and reserpine are mediated by the neural input to the medial basal hypothalamus, whereas those of hypoxia, insulin hypoglycemia, histamine, *E. coli* endotoxin, 5-hydroxytryptophan and a large dose of formaldehyde depend predominantly on humoral mechanism. It is also suggested that both neural and humoral mechanisms might participate in pituitary-adrenocortical activation in rats induced by tourniquet, cold and heat exposure. As to the stimulatory effect of ether anesthesia, however, experimental results are rather controversial. The findings obtained in experiments on rats with various partial hypothalamic deafferentations suggest that the pituitary-adrenocortical secretory responses to reserpine, sciatic nerve stimulation, olfactory stimulation and leg break depend predominantly on neural impulses to the medial basal hypothalamus through the anterior afferent pathway, those to exposure to noise and vibration through the anterior, lateral and dorsal afferent pathways, and those to audiogenic and photic stimulations and exposure to heat through the posterior and/or lateral afferent pathways.

In pentobarbital-anesthetized dogs *Lau* [1971b] has evaluated the effect of hypothalamic deafferentation on adrenal 17-OHCS secretion in response to hypoxia and hypercapnia. In her experiments an increase in adrenal 17-OHCS secretion rate during exposure to 10% O_2 in N_2 and 10% O_2 + 5% CO_2 in N_2 was found to be prevented by complete deafferentation of the hypothalamus. However, a significant elevation of adrenal 17-OHCS secretion rate in dogs with complete hypothalamic deafferentation was observed during breathing 10% O_2 + 10% CO_2 in N_2. These findings suggest that hypoxic hypoxia accompanied with hypocapnia or normocapnia stimulates the pituitary-adrenocortical system via neural pathway and hypoxia accompanied with elevated arterial CO_2 tension may directly stimulate the hypothalamic-pituitary-adrenocortical system without neural input to the hypothalamus.

Table VI. Effect of complete deafferentation of the medial basal hypothalamus in rats on plasma cortico-sterone responses to various stressful stimuli

References	Stressful stimulus	Plasma corticosterone response
Feldman et al. [1968]	ether	unaffected
	audiogenic stimulation[1]	completely abolished
Voloschin et al. [1968]	ether	completely abolished
Palka et al. [1969]	ether	unaffected
	immobilization	significantly reduced but not completely abolished
Greer et al. [1970]	leg break	completely abolished
	tourniquet	reduced but not completely abolished
Feldman et al. [1970]	immobilization	unaffected
	hypoxia (exposure to N_2)	unaffected
	ether	unaffected
	audiogenic stimulation[1]	abolished
	optic stimulation[2]	significantly reduced
Makara et al. [1970]	*E. coli* endotoxin (25 μg/100 g, i.p.)	unaffected
	histamine (5 mg/100 g, i.p.)	unaffected
	insulin (0.25 U/100 g, i.p.)	unaffected
	a large dose of formaldehyde (25 mg/100 g, s.c.)	unaffected
	capsaicin (0.25 mg/100 g, s.c.)	abolished
Chowers et al. [1972]	exposure to cold (0° C, 30 min)	significantly reduced
Lengvári and Halász [1972]	reserpine (3.0–3.5 mg/kg, i.p.)	completely abolished
Popova et al. [1972]	5-hydroxytryptophan (100 mg/kg, i.p.)	unaffected
Feldman et al. [1975a]	ether	unaffected
	sciatic nerve stimulation (2 min)	completely abolished
Chowers et al. [1976]	exposure to heat (36 °C, 30 min)	significantly reduced
	fifth 30-min exposure to heat (36 °C)[3]	significantly reduced
	exposure to heat (36 °C, 120 min)	significantly reduced
	fifth 120-min exposure to heat (36 °C)[3]	unaffected
Feldman and Conforti [1976b]	ether	unaffected
	olfactory stimulation[4]	almost completely abolished
Makara et al. [1980]	ether	completely abolished

[1] Exposure for 30 min to the sound of an alarm clock.
[2] Exposure for 30 min to a Grass photostimulator emitting flashes (5/s).
[3] Heat-exposure was repeated 5 times with 48-hour intervals between each experiment.
[4] Exposure for 30 min to the air containing amyl acetate.

Table VII. Effect of various partial deafferentations of the medial basal hypothalamus in rats on plasma corticosterone responses to various stressful stimuli

References	Location of partial deafferentation	Stressful stimulus	Plasma corticosterone response
Makara et al. [1969]	anterolateral	exposure to noise and vibration	abolished
		sham adrex[1]	abolished
		formalin (1%, 0.25 ml/100 g, s.c.)	abolished
		histamine (1 mg/100 g, i.p.)	significantly reduced
		capsaicin (0.25 mg/100 mg, s.c.)	significantly reduced
		E. coli endotoxin (25 µg/100 g, i.p.)	unaffected
Greer et al. [1970]	anterior quarter	leg break (ipsilat.)	unaffected
		leg break (contralat.)	significantly reduced
	posterior quarter	leg break (ipsilat.)	unaffected
		leg break (contralat.)	unaffected
Chowers et al. [1972]	anterior	exposure to cold (0 °C, 30 min)	unaffected
	posterolateral	exposure to cold (0 °C, 30 min)	significantly reduced
	posterior	exposure to cold (0 °C, 30 min)	significantly reduced
Feldman et al. [1972a]	anterior	ether	unaffected
		audiogenic stimulation[2]	unaffected
		photic stimulation[3]	unaffected
	anterolateral (or posterolat.)	ether	unaffected
		audiogenic stimulation[2]	significantly reduced
		photic stimulation[3]	significantly reduced
Feldman et al. [1972b]	posterior	ether	slightly reduced
		audiogenic stimulation[2]	markedly reduced
		photic stimulation[3]	markedly reduced
Lengvári and Halász [1972]	anterior	reserpine (3.0–3.5 mg/kg, i.p.)	abolished
	posterior	reserpine (3.0–3.5 mg/kg, i.p.)	significantly reduced but not completely blocked
Makara et al. [1972]	anterior	sham adrex[1]	unaffected
	'low' superior	sham adrex	unaffected
	'low' anterosup.	sham adrex	unaffected
	'short' posterior	sham adrex	unaffected
	'high' superior	sham adrex	significantly reduced
	'high' anterosup.	sham adrex	significantly reduced
	'long' posterior	sham adrex	significantly reduced
	parasagittal (lat.)	sham adrex	abolished
Feldman et al. [1975b]	anterior	ether	unaffected
	(or anterolat.)	sciatic nerve stimulation	significantly reduced
	posterior	ether	unaffected
	(or posterolat.)	sciatic nerve stimulation	unaffected

Table VII. (continued)

References	Location of partial deafferentation	Stressful stimulus	Plasma corticosterone response
Chowers et al. [1976]	anterior	exposure to heat (36 °C, 30 min)	unaffected
		exposure to heat (36 °C, 120 min)	unaffected
	posterior	exposure to heat (36 °C, 30 min)	significantly reduced
		exposure to heat (36 °C, 120 min)	unaffected
Feldman and Conforti [1977]	anterior	ether	unaffected
		olfactory stimulation[4]	completely abolished
	posterior	ether	unaffected
		olfactory stimulation	unaffected
Makara et al. [1980]	anterolateral	ether	completely abolished

[1] Sham adrenalectomy.
[2] Exposure for 30 min to the sound of an alarm clock.
[3] Exposure for 30 min to a Grass photostimulator emitting flashes (5/s).
[4] Exposure for 30 min to the air containing amyl acetate.

The effect of hypothalamic deafferentation in the rat on the circadian periodicity of plasma corticosterone levels has been studied by some investigators.

Halász et al. [1967] have shown that the a.m. resting levels of plasma corticosterone in rats with complete deafferentation of the medial basal hypothalamus are elevated when compared with those in control rats; no further increase or decrease in plasma concentration of corticosterone has been observed in the evening. Thus, the circadian rhythmicity of plasma corticosterone levels in rats has been found to be abolished by complete hypothalamic deafferentation. The simplest interpretation of this finding would be that the effect of complete hypothalamic deafferentation on basal levels of plasma corticosterone in rats is due to removal of inhibiting influences on the pituitary-adrenocortical secretory activity. In contrast, *Palka* et al. [1969] have suggested that the main effect of complete deafferentation of the medial basal hypothalamus in the rat on the pituitary-adrenocortical system is due to removal of nervous impulses to the medial basal hypothalamus which produce the p.m. elevation of plasma corticosterone levels in intact animals, since they have found that the p.m. levels of plasma corticosterone in hypothalamic-deafferented rats are significantly lower than those in control rats and the a.m.–p.m. difference in plasma corticosterone levels, which is observed in the latter group of rats, is absent in the former group. Similar findings were reported by *Fukuda and Greer* [1975]. *Stark* et al. [1978] failed to find any elevation of a.m. basal plasma levels of corticosterone in rats induced by complete deafferentation of the medial basal hypothalamus.

Wilson and Critchlow [1975] determined basal plasma levels of corticosterone of peripheral blood samples obtained from each rat at 4-hour intervals for 44 h. In most intact and sham-operated rats the peak corticosterone levels were observed at the time of light-dark transition (1800 hours). In rats with complete deafferentation of the medial basal hypothalamus, each animal showed marked fluctuations in plasma corticosterone levels. The mean values of the maximal variations and of peaks in hypothalamic-deafferented rats were much the same with those in controls. However, the fluctuations in each rat were quite asynchronous and there was no clear circadian rhythm; often more than one peak and sometimes no clear peaks were observed during a 24-hour period. The experimental results might indicate that the medial basal hypothalamus stimulates episodically the pituitary-adrenocortical secretory activity, but cannot stimulate the latter to produce circadian rhythm, without neural input from other brain structures.

Lengvári and Liposits [1977b] reported that the circadian rhythmicity of plasma corticosterone levels in rats, which had been abolished following frontal deafferentation of the medial basal hypothalamus, was found to reappear by the 36th week after deafferentation.

The circadian periodicity of peripheral plasma levels of cortisol in rhesus monkeys has been found to be abolished by complete deafferentation [*Krey* et al., 1975; *Ferin* et al., 1977; *Spies* et al., 1979] but not by anterior [*Krey* et al., 1975] or posterior [*Spies* et al., 1979] partial deafferentation of the medial basal hypothalamus.

E. Hypothalamic Corticotropin-Releasing Factor

Evidence has been accumulating of the control of the pituitary ACTH release by a substance, corticotropin-releasing factor (CRF) or corticotropin-releasing hormone (CRH), produced in the hypothalamus and conveyed to the anterior pituitary by the hypophyseal portal circulation. Purification and identification of CRF from hypothalamic tissue extract have been attempted by a number of investigators and various in vitro and in vivo methods for estimating CRF activity of the test materials have been devised.

Suggestive evidence has been offered by *Porter and Jones* [1956] and *Porter and Rumsfeld* [1956, 1959] in favor of the view that CRF is released into the hypophyseal portal vessels. In their experiments chloralose-urethane-anesthetized dogs, whose pituitaries had been exposed, were injected with heparin and subjected to hypophysectomy (removal of the pituitary by aspiration). The blood from the broken hypophyseal portal vessels was collected and plasma of the blood sample was evaluated for CRF activity. Injection (i.v.) of the plasma sample into the cortisol-treated intact rats, but not into the hypophysectomized rats, was found to produce a significant depletion of adrenal ascorbic acid. However, the specificity of the method used in their studies for assaying CRF activity has been questioned by *Fortier* [1966].

Hypothalamic CRF Activity following Stress

A rapid and significant elevation of CRF activity of the rat median eminence after stress (exposure to ether vapor for 1 min followed by sham unilateral adrenalectomy) has been demonstrated by *Vernikos-Danellis* [1964]. *Hiroshige* et al. [1969c] have shown a significant increase in CRF activity of the rat median eminence following systemic administration of epinephrine, histamine or arginine vasopressin, exposure to ether vapor followed by laparotomy and exposure to formalin gas or noise (bell-ringing), an increase which was observed 2–8 min after the start of exposure to stress. *Takebe* et al. [1971a] also observed a significant increase in median eminence CRF activity in rats 2 min after the start of exposure to stressful stimuli (laparotomy followed by intestinal traction).

Hiroshige et al. [1971] studied the time course of changes in CRF activity of the rat median eminence following exposure to ether vapor (followed by laparotomy). The median eminence CRF activity increased rapidly and peaked at 2 min after the start of ether-laparotomy stress. It subsequently decreased to the level below the baseline by 20 min. Then, it increased again and attained a second peak at 80 min. Thereafter it decreased and was markedly depleted at 100 min. Similar changes with 2 peaks in CRF activity of the rat median eminence following surgical stress were reported by *Sato* et al. [1975] and *Fujieda and Hiroshige* [1978]. The latter investigators showed that pretreatment with cycloheximide, an inhibitor of de novo synthesis of protein, abolished the second but not the first peak of the hypothalamic CRF activity in rats following surgical stress.

A significant increase in CRF activity of the SME in rats 1.5 min after the start of audiogenic stress [*Chowers* et al., 1973] and that of the median eminence in rats just after a 2-min immobilization [*Sakakura* et al., 1976a,b] have also been noted. *Chowers* et al. [1969] reported that the CRF activity of the SME in rats was significantly elevated following 7-day starvation. *Karaulova* [1973] evaluated the changes in hypothalamic CRF activity in rats following muscular exercise. The CRF activity was found to be significantly increased after 30-min swimming and significantly decreased following swimming for 100–110 min. A significant decrease in hypothalamic CRF in rats after exposure to heat for 10–15 (or 15–20) min has been observed [*Degonskii*, 1973, 1977].

Circadian Variation of the Hypothalamic CRF Activity
David-Nelson and Brodish [1969] have shown that hypothalamic CRF activity in the rat gradually increases from early morning hours through mid-afternoon, attains the peak level at 1500 hours, suddenly drops at 1800 hours, rapidly returns at 2100 hours and then gradually decreases until early morning hours. The CRF activity of the rat median eminence has been found to be lower in the morning than in the evening [*Hiroshige* et al., 1969a; *Hiroshige and Sato*, 1970a,b; *Takebe and Sakakura*, 1972]. However, *Retiene and Schulz* [1970] and *Sirett and Purves* [1973] found in the rat only a slight or no change of hypothalamic CRF activity.

In intact male rats, the CRF activity of the median eminence and plasma corticosterone content are similar in their patterns of circadian variations, though the changes in the former mostly precede those in the latter [*Hiroshige* et al., 1969a; *Hiroshige and Sakakura*, 1971; *Takebe* et al., 1971b, 1972; *Sakakura* et al., 1978].

In adrenalectomized male rats, the hypothalamic CRF activity was found by *Cheifetz* et al. [1969] to be significantly higher in the morning than in the evening, whereas it was shown by *Hiroshige and Sakakura* [1971] to be lower in the morning and peaked in the afternoon at the time about 2 h earlier than in intact male rats.

The sex difference in circadian variations of hypothalamic CRF activity in rats has been studied by *Hiroshige* et al. [1973]. They have observed that the CRF activity in intact female rats is higher in the morning than in the afternoon. Subsequently, *Hiroshige and Wada-Okada* [1973] reported that the CRF activity of the medial basal hypothalamus in female rats during diestrus did not show any significant circadian variations, whereas during proestrus and estrus the CRF activity was higher in the morning than in the afternoon. *Szafarczyk* et al. [1980b] noted the circadian periodicities of hypothalamic CRF activity and plasma corticosterone levels with peaks appearing in the evening in ovariectomized estradiol-implanted female rats.

Vasopressin and Hypothalamic CRF

It has been shown that vasopressin has a CRF-like action, i.e. it stimulates directly the anterior pituitary to release ACTH. However, vasopressin and hypothalamic CRF are not identical, since the dose-response curve of vasopressin in the in vitro tests is non-parallel with that of hypothalamic extract preparation containing CRF [*Portanova and Sayers,* 1973a; *Fehm* et al., 1975; *Buckingham and Hodges,* 1977; *Krieger* et al., 1977b; *Lutz-Bucher* et al., 1977; *Seelig and Sayers,* 1977; *Gillies* et al., 1978; *Gillies and Lowry,* 1978, 1979, 1980]. *Gillies and Lowry* [1979] have postulated that hypothalamic CRF is vasopressin modulated by some hypothalamic factor(s).

Yates et al. [1971] showed that the pretreatment with vasopressin (i.v. or bilaterally into the anterior pituitary in subthreshold doses for stimulating the pituitary ACTH release) in the rat markedly potentiated the plasma corticosterone elevating action of crude median eminence extract (i.v.) but not of ACTH (i.v.) . The findings may be interpreted to indicate that vasopressin potentiates the action of hypothalamic CRF at the pituitary level. In the in vitro experiments with incubated fragments of rat anterior pituitary *Lutz-Bucher* et al. [1980] found that ACTH-releasing effects of vasopressin and a maximum effective dose of hypothalamic CRF (vasopressin-free crude median eminence extract) were additive at the pituitary level, suggesting that vasopressin has in the anterior pituitary a receptor site different from that of CRF.

The effect of removal of the posterior pituitary (neurohypophysectomy) on the pituitary-adrenocortical responses in rats to a variety of stresses has been studied by *Nowell* [1959], *Fisher and De Salva* [1959], *Smelik* [1960], *Arimura and Long* [1962], *Itoh* et al. [1964], *Arimura* et al.

[1965], *De Wied* [1968] and *De Wied* et al. [1969]. Experimental results of the above investigators suggest that the posterior lobe of the pituitary is not essential in the control of the pituitary-adrenocortical activity, though it might play some role in adrenocortical responses to certain stresses.

The physiological role of vasopressin in the control of the pituitary-adrenocortical secretory activity has been studied using congenitally vasopressin-deficient rats (Brattleboro rats). *McCann* et al. [1966] have observed that the elevation of plasma corticosterone levels in response to ether stress is significantly reduced, though it is not completely abolished, in Brattleboro rats. They have also observed that plasma corticosterone response to bleeding under mild restraint and adrenal ascorbic acid depletion in response to unilateral adrenalectomy are partially impaired in Brattleboro rats. The elevation of plasma corticosterone levels in response to a variety of stresses, such as histamine (i.v.) or laparotomy followed by intestinal traction after pretreatment of the animals with dexamethasone, exposure to noise or ether vapor [*Yates* et al., 1971], histamine (i.v.), cage moving with or without bell ringing, ether-laparotomy-hemorrhage [*Wiley* et al., 1974] has been found to be significantly depressed in Brattleboro rats. In contrast, *Arimura* et al. [1967] showed that plasma corticosterone responses to ether, histamine, vasopressin or acetylcholine in Brattleboro rats were not significantly different from those in normal rats, although the elevation of plasma corticosterone levels following i.p. injection of 0.9% saline was significantly depressed in the former rats.

Arimura et al. [1967], *Krieger* et al. [1977b] and *Gillies and Lowry* [1980] found that the hypothalamic CRF activity in Brattleboro rats was significantly less than in normal rats, whereas *McCann* et al. [1966] failed to find any significant difference in hypothalamic CRF activity between Brattleboro rats and normal rats.

Chapter 7. Participation of the Extrahypothalamic Brain Structures and Spinal Cord in the Control of Pituitary-Adrenocortical Activity

A. Amygdaloid Stimulation

Adrenal ascorbic acid depletion in rats following electrical stimulation of the amygdaloid nucleus has been reported [*Endröczi* et al., 1959].

The effect of electrical stimulation of various areas of the limbic system in the dog on pituitary-adrenocortical activity has been studied by *Okinaka* et al. [1960a] and *Okinaka* [1961]. They found that stimulation of the amygdala in dogs anesthetized with morphine (ether was used when necessary) resulted in a significant elevation of adrenal 17-OHCS secretion. *Slusher and Hyde* [1961b] reported that in encéphale isolé cat preparations a significant increase in corticosteroid levels of the adrenal venous effluent was produced by stimulation of the medial part of the basal nucleus of the amygdala, a significant decrease by stimulation of the lateral amygdaloid nucleus or lateral part of the basal amygdaloid nucleus and no significant change by stimulation of the amygdala on the border between the basomedial and basolateral nuclei. In pentobarbital-anesthetized dogs *Ishihara* et al. [1964] observed an increase in adrenal 17-OHCS secretion following electrical stimulation of the amygdala. An increase in adrenal 11-OHCS secretion rate in pentobarbital-anesthetized dogs elicited by electrical stimulation of the medial and lateral nuclei of the amygdala and a decrease by stimulation of the lateral part of the basal amygdaloid nucleus have also been reported [*Ishihara* et al., 1965b].

Mason [1959] found a marked elevation of peripheral plasma 17-OHCS levels in conscious rhesus monkeys following electrical stimulation of the amygdala, an elevation comparable to that occurring after hypothalamic stimulation or administration of a large dose of ACTH. In his paper it was also noted that in 1 monkey anesthetized with pentobarbital and in another monkey, whose distal hippocampus had been stimulated 48 h previously, stimulation of the amygdala failed to elevate peripheral plasma levels of 17-OHCS.

A rise of plasma corticosteroid levels following electrical stimulation of the amygdala has been confirmed in cats during recovery stage of a long-lasting dial anesthesia [*Setekleiv* et al., 1961], in conscious cats [*Matheson* et al., 1971], in pentobarbital-anesthetized dogs [*Salcman* et al., 1970], in pentobarbital-anesthetized rats [*Redgate and Fahringer*, 1973], in conscious rhesus monkeys [*McHugh and Smith*, 1967; *Ehle* et al., 1977], in phencyclidine-anesthetized rhesus monkeys [*Frankel* et al., 1978] and in man [*Mandell* et al., 1963; *Rubin* et al., 1966]. In rhesus monkeys, electrical stimulation of the amygdala which evokes amygdaloid after-discharges consistently produces an elevation of peripheral plasma levels of 17-OHCS, whereas the stimulation which is not associated with amygdaloid after-discharges only produces small and insignificant changes in plasma 17-OHCS levels [*McHugh and Smith*, 1967]. The elevation of peripheral plasma levels of cortisol and corticosterone in cats is induced by electrical stimulation of the corticomedial, basal and lateral portions, but not of the anterior portion, of the amygdala [*Matheson* et al., 1971]. Plasma cortisol levels in the monkey are significantly elevated by electrical stimulation of the basal and lateral portions but not the corticomedial part of the amygdala [*Frankel* et al., 1978].

In conscious rabbits electrical stimulation of the amygdala has been shown to increase the biosynthetic activity of the adrenal cortex [*Kawakami* et al., 1968a].

B. Hippocampal Stimulation

Endröczi et al. [1959] failed to find any significant change in adrenal ascorbic acid content in rats following electrical stimulation of the gyrus or uncus hippocampi. However, they demonstrated that adrenal ascorbic acid depletion in rats in response to painful electrical stimulation of the hind limb or to epinephrine (s.c.) was abolished by simultaneous stimulation of the hippocampus.

Okinaka et al. [1960a] and *Okinaka* [1961] showed that electrical stimulation of the hippocampus resulted in a slight decrease in adrenal 17-OHCS secretion rate. *Slusher and Hyde* [1961b] observed in encéphale isolé cat preparations a significant decrease in adrenal venous ef-

fluent corticosteroid levels following electrical stimulation of the hippocampal uncus. In 1 cat, however, they could find a significant
increase following stimulation of the anterodorsal area of the dentate
gyrus of the hippocampus.

The effect of hippocampal stimulation on adrenocortical secretion
has been found to depend on the stimulatory parameters [*Endröczi and
Lissák*, 1962]. In pentobarbital-anesthetized cats electrical stimulation
of the dorsal hippocampus at a frequency of 12 or 36 cps markedly inhibits the adrenocortical secretion in response to painful stimuli (electrical stimulation of the legs), whereas hippocampal stimulation at a
frequency of 240 cps enhances the response.

In man electrical stimulation of the hippocampus produces a depression of peripheral plasma 17-OHCS levels [*Mandell* et al., 1963;
Rubin et al., 1966], whereas in phencyclidine-anesthetized monkeys
stimulation of the anterior part of the hippocampus results in an increase in peripheral plasma cortisol [*Frankel* et al., 1978]. The effect of
electrical hippocampal stimulation on plasma corticosterone levels in
non-stressed rats has been extensively studied by *Casady and Taylor*
[1976]. Stimulation of the ventral hippocampus, performed in the afternoon, induced a significant increase followed by a significant decrease
in plasma corticosterone levels. Stimulation in the afternoon of the
dorsal hippocampal area resulted in a significant increase followed by
a less pronounced decrease. Morning stimulation of the dorsal hippocampal area elicited a significant increase, which was not followed by a
decrease, in plasma corticosterone concentrations.

In the rabbit, *Kawakami* et al. [1968a] have demonstrated that the
accelerated adrenal corticosteroid biosynthesis in response to immobilization stress is inhibited by electrical stimulation of the hippocampus
which under non-stress conditions accelerates the adrenocortical biosynthesis.

C. Septal, Cingulate and Orbital Cortical Stimulation

In experiments with encéphale isolé cat preparations *Slusher and
Hyde* [1961b] have found that electrical stimulation of the septum or
the diagonal band of Broca produces a significant depression of corticosteroid levels of adrenal venous effluent. In pentobarbital-anesthetized cats *Endröczi and Lissák* [1963] have observed a marked decrease

in adrenocortical secretion following electrical stimulation of the basal septum. *Okinaka* et al. [1960b] have shown that electrical stimulation of the posterior orbital surface in morphine-anesthetized dogs induces a prompt and transient increase in adrenal venous blood concentrations of 17-OHCS, a response considerably delayed by bilateral splanchnic nerve section and abolished by hypophysectomy. Stimulation of the anterior Sylvian gyrus has no effect.

Setekleiv et al. [1961] evaluated the effect of cerebral cortical stimulation in cats during recovery stage of a long-lasting dial anesthesia on plasma 17-OHCS levels. Peripheral plasma levels of 17-OHCS were found to be elevated by stimulation of the anterior cingulate cortex and the lower part of the posterior ecto- and supra-Sylvian gyri. In phencyclidine-anesthetized monkeys the plasma cortisol response to electrical stimulation of various areas of the limbic system has been extensively studied by *Frankel* et al. [1978]. Stimulation of the anterior part of the cingulate gyrus close to the corpus callosum, the paracingulate area and the grey matter of the orbital cortex of the frontal lobe elicited a definite elevation of peripheral plasma cortisol levels. Stimulation of the piriform cortex resulted in a late elevation, whereas stimulation of the septum had no effect.

D. Brain Stem Stimulation

In experiments with encéphale isolé cat preparations *Slusher and Hyde* [1961a] have shown a significant decrease in corticosteroid secretion induced by electrical stimulation of the ventral midbrain tegmentum. In contrast, *Endröczi and Lissák* [1963] observed in pentobarbital-anesthetized cats a marked increase in adrenocortical secretion in response to electrical stimulation of the ventral tegmentum and a significant decrease following stimulation of the dorsal tegmentum of the midbrain.

As to the effect of electrical stimulation of the mesencephalic reticular formation on the pituitary-adrenocortical secretory activity, *Okinaka* et al. [1960a] and *Okinaka* [1961] have found that adrenal 17-OHCS secretion in morphine-anesthetized dogs is moderately elevated by stimulation of the caudal, but not the rostral, portion of the mesencephalic reticular formation. In pentobarbital-anesthetized cats, however, *Endröczi and Lissák* [1963] showed that a significant increase in adrenocortical secretion was induced by electrical stimulation of the mesencephalic reticular formation at the level of, or rostral and ventral to, the superior colliculus.

E. Brain Removal and Decortication

Adrenal 17-OHCS secretion in response to surgical stress in pento-barbital-anesthetized dogs following stepwise removal of the brain has been evaluated by *Story* et al. [1959]. They found that the elevated adrenal 17-OHCS secretion induced by surgical stress was not depressed by the removal of the brain down to the hypothalamus.

An elevated resting 17-OHCS secretion has been observed in unanesthetized dogs whose whole brain (including the hypothalamus) down to the inferior colliculus had been removed leaving the pituitary intact, and in two thirds of these dogs a further increase in adrenal 17-OHCS secretion has been produced by burn of the paw and lower foot [*Egdahl*, 1960a] and by sciatic nerve stimulation [*Egdahl*, 1960b]. *Egdahl* [1960a] postulated that in dogs with brain removal and isolated pituitaries a neurohormone might be released from the hindbrain (hindbrain factor, HBF) autonomously or as a response to stressful stimuli and transported by systemic circulation to the anterior pituitary to stimulate ACTH release; in intact dogs the release of HBF might be tonically inhibited by higher brain areas. A significant increase in adrenal 17-OHCS secretion in pentobarbital-anesthetized brain-removed dogs following partial constriction of the thoracic inferior vena cava has also been demonstrated [*Egdahl*, 1961a].

However, *Wise* et al. [1962, 1963] failed to support the above hypothesis of *Egdahl* [1960a]. They evaluated the adrenal 17-OHCS secretion in response to surgical stress of adrenal vein cannulation in dogs whose whole brain rostral to the pons had been removed, leaving the hindbrain intact, and in dogs whose entire brain (including the hindbrain) down to the level of the first cervical spinal segment had been extirpated. It was found that adrenal 17-OHCS secretion in the latter group of dogs was not significantly different from that of the former group of dogs. Similar findings were also obtained by *Egdahl* [1962]. In pentobarbital-anesthetized dogs with removal of the brain down to the pons leaving the pituitaries intact, adrenal 17-OHCS secretion was first depressed during several hours following brain removal, then a sustained marked elevation of 17-OHCS secretion was observed, which was not altered by additional removal of the hindbrain and even by removal of the spinal cord down to the cauda equina in addition to the entire brain.

Egdahl [1962] noted that an elevated adrenocortical secretion in dogs with brain removal and isolated pituitaries was not reduced by additional removal of the posterior pituitary. However, *Wise* et al. [1963] pointed out the difficulty of complete removal of the hypothalamus and neurohypophysis in the dog without severely damaging the adenohypophysis, and they suggested the possibility that an increased adrenocortical secretion in dogs with isolated pituitaries might be mediated by CRF released from degenerating neural elements such as neurohypophyseal or hypothalamic remnants.

Kendall and Roth [1969] demonstrated in rhesus monkeys with removal of the forebrain and intact pituitaries a persistent increase in adrenal 11-OHCS secretion similar to that in brain-removed dogs. As a possible explanation of this finding, they suggested the supersensitivity of the anterior pituitary after isolation from the brain analogous to 'denervation supersensitivity'.

The effect of brain removal on adrenocortical secretory activity in rats has been found to be quite different from that in dogs. Plasma corticosterone levels in rats have never been persistently elevated after removal of the forebrain rostral to the superior colliculus leaving isolated pituitaries [*Matsuda* et al., 1963; *Dunn and Critchlow,* 1969].

The effect of extirpation of cerebral cortices down to the hippocampus on adrenal 17-OHCS secretion in dogs has been studied by *Egdahl* [1961b]. Adrenal 17-OHCS secretion rates in unanesthetized dogs, which were 1.0 ± 0.2 µg/min before decortication, were elevated on the following day after decortication to 10.2 ± 0.8 µg/min. Following i.v. injection of sodium pentobarbital the secretion rates were promptly decreased to 2.1 ± 0.8 µg/min, which were again elevated to 14.1 ± 2.1 µg/min by electrical stimulation of the sciatic nerve. It is suggested that the release of hypothalamic CRF is tonically stimulated by nervous impulses from the reticular formation and the latter might be constantly inhibited by impulses from the cerebral cortices. Pentobarbital anesthesia might depress the reticular formation but not the hypothalamus, since a marked elevation of adrenocortical secretion could be produced by stimulation of the sciatic nerve in pentobarbital-anesthetized decorticated dogs.

F. Amygdaloid Lesions

Martin et al. [1958] have found that in dial-anesthetized dogs and cats a marked increase in adrenal secretion of cortisol and corticosterone is produced by complete removal of the amygdala and periamygdalic cortical areas. In the deermouse significant increases in plasma and adrenal corticosterone have been observed following lesions of the medial amygdaloid nuclei [*Eleftheriou* et al., 1966]. In contrast, *Knigge* [1961] and *Seggie* [1980] failed to find in rats any significant effect of bilateral destruction of the amygdala on resting levels of plasma corticosterone. The latter investigator also noted that the circadian variations of plasma corticosterone levels in rats were not significantly altered by bilateral amygdala lesions.

Ishihara et al. [1964] noted that the removal of the amygdala in pentobarbital-anesthetized dogs could not suppress the adrenal 17-OHCS secretion in response to surgical stress (laparotomy and adrenal vein cannulation), electrical stimulation of the sciatic nerve and histamine (s.c.). However, they showed that the peripheral plasma 11-OHCS response to immobilization in conscious dogs was partially suppressed by removal of the amygdala.

The effect of amygdala lesions on plasma corticosterone response to continuous immobilization in rats has been evaluated by *Knigge* [1961]. He observed that in rats with amygdala lesions the elevation of plasma corticosterone levels in response to immobilization was delayed in reaching the maximal levels. Bilateral lesions of the amygdala in rats have been found to suppress or block the plasma corticosterone response to etherization and withdrawal of blood sample by heart puncture [*Knigge and Hays,* 1963]. *Allen and Allen* [1974] have shown that bilateral amygdaloid lesions in rats block the elevation of plasma corticosterone levels in response to a leg break, but not to a tourniquet or ether stress. They have also shown that bilateral lesions of the areas adjacent to the amygdala and lateral hypothalamus, involving a direct medial projection of the amygdala to the hypothalamus, also block the elevation of plasma corticosterone levels in response to hind-leg break but not to etherization or hind-leg tourniquet. These findings suggest that the amygdaloid nuclei facilitate the effect of neurogenic stress, such as hind-leg break, on the pituitary-adrenocortical secretory activity and the pathway for this facilitatory effect might traverse the direct medial amygdaloid projection to the hypothalamus.

G. Hippocampal Lesions

A sustained inhibitory effect of the hippocampus on the pituitary-adrenocortical activity has been demonstrated by *Kim and Kim* [1961] and *Knigge* [1961]. The former investigators observed that adrenal ascorbic acid depletion in rats in response to acute stress (ether anesthesia followed by unilateral adrenalectomy) was enhanced by lesions of the hippocampus. The latter investigator found that the resting plasma corticosterone levels in the rat were significantly elevated after hippocampal lesions; the mean values in rats with hippocampal lesions and in normal rats were 20.6 ± 3.8 (SD) and 11.8 ± 2.1 µg/100 ml, respectively.

Knigge and Hays [1963] have shown that the elevation of plasma corticosterone levels in rats in response to etherization followed by heart puncture is not impaired by bilateral hippocampal lesions but it is prevented by bilateral lesions of the amygdala or midbrain reticular formation, and that the block produced by either of the latter two lesions is eliminated by placement of additional bilateral lesions of the hippocampus. *Coover* et al. [1971] failed to find any significant effect of hippocampal lesions in rats on basal levels and the response to ether stress (etherization plus venesection), exposure to a new environment and water deprivation, of plasma corticosterone. However, they found that in rats with hippocampal lesions plasma corticosterone levels during a reinforcement session were markedly depressed and the extinction session elevation was abolished. *Wilson and Critchlow* [1973/74] reported that the circadian rhythm of non-stress plasma corticosterone levels remained unimpaired and ether-induced elevation of plasma corticosterone levels was significantly reduced in hippocampectomized rats.

Lanier et al. [1975] have studied the effects of differential hippocampal damage (dorsal, ventral, and near-total lesions of the hippocampus) in rats on the circadian variation and ether-induced elevation of plasma corticosterone levels. They failed to find any significant alteration in the response to ether stress and circadian variations of plasma corticosterone levels.

H. Fornix Transection

Moberg et al. [1971] have demonstrated that the circadian rhythm of plasma corticosterone levels in rats is abolished 1–2 weeks after bilateral section of the fornix. In normal rats the levels of plasma corticosterone are the lowest at 0800 hours and the highest at 2000 hours. After bilateral section of the fornix with or without damage of the stria terminalis the levels of plasma corticosterone are intermediate between the diurnal low and high values in normal rats throughout the day, i.e. significantly higher at 0800 hours and significantly lower at 2000 hours than those in normal controls. *Lengvári and Halász* [1973] also examined the plasma corticosterone levels in rats 1 and 3 weeks after fornix transection. 1 week after section of the fornix no circadian rhythm of plasma corticosterone was observed. In rats 3 weeks after bilateral section of the fornix, however, normal diurnal variations of plasma corticosterone levels were observed. They suggested that the neural input from the hippocampus to the hypothalamus which is mediated through the fornix might not be essential for the circadian rhythm of pituitary-adrenocortical activity, or it could easily be replaced by other brain structures. *Wilson and Critchlow* [1973/74] noted a normal pattern of circadian rhythm in plasma corticosterone levels in rats 8–11 days and 29–32 days after fornix section. They also noted a significant plasma corticosterone response to ether in fornix-sectioned rats, though it was reduced at 8–11 days but not 29–32 days after fornix section in comparison to that in control rats. The above findings suggest that the neural pathway through the fornix is not essential for the circadian rhythm and response to ether stress of the pituitary-adrenocortical system.

I. Septal and Cingulate Lesions

Baseline plasma levels of corticosterone in rats are unaffected by septal lesions [*Usher* et al., 1967; *Endröczi and Nyakas*, 1971; *Uhlir* et al., 1974; *Wilson and Critchlow*, 1974]. *Usher* et al. [1967] have shown that the elevation of plasma corticosterone levels in rats in response to chronic stress (1-second blasts of compressed air at 5- to 10-min intervals for the entire 3-day period) is abolished by lesions of the lateral septal area or the anterior cingulate gyrus. In contrast, it has been noted that the plasma corticosterone response in rats to passive avoidance behavior testing is significantly enhanced by septal lesions [*Endröczi and Nyakas*, 1971] and the threshold level of stimulus (electric shock) for elevation of plasma corticosterone levels is lower in rats with septal lesions than in control rats [*Uhlir* et al., 1974].

Seggie and Brown [1971] reported that resting plasma corticosterone levels at a time in the light-dark cycle 2 h after lights on in septum-lesioned rats showed a 5-fold increase over the values in control rats, though the circadian variations of plasma corticosterone still persisted after septal lesions. The experimental results suggest that the circadian rhythm of plasma corticosterone in rats could be altered by septal lesions. By contrast, *Wilson and Critchlow* [1974] observed normal patterns of the circadian rhythm in plasma corticosterone levels in rats 3, 5 and 9 weeks after removal of the septum with concomitant section of the fornix. The findings may be interpreted to indicate that the septum and fornix are not essential for the circadian rhythm of pituitary-adrenocortical activity.

Seggie et al. [1974] have examined the effect of septal lesions on the circadian rhythm of plasma corticosterone levels in rats under carefully controlled (rigorous) environmental conditions and usual (standard) conditions. In septally lesioned rats maintained under rigorous environmental conditions, plasma corticosterone levels, which were measured at 8 points of time during 24 h, were found at each point to be identical to those seen in control rats under the same environmental conditions. However, when plasma corticosterone levels in septum-lesioned rats kept under standard environmental conditions were measured at 6 points of time during 24 h, they were significantly higher at a time of the light-dark cycle 2 h before lights on than those in control rats under the same environmental conditions. The above findings suggest that the circadian rhythm of plasma corticosterone in rats is unaltered after septal lesions and the responsiveness of the pituitary-adrenocortical system to various stimuli such as environmental disturbances would be enhanced by lesions of the septum.

J. Brain Stem Transection and Lesions

Martini et al. [1960] and *Giuliani* et al. [1961] have shown that adrenal ascorbic acid depletion in the rat in response to stressful stimuli, such as exposure to ether followed by unilateral adrenalectomy, is completely abolished by midbrain transection. The latter investigators have also demonstrated that adrenal ascorbic acid responses in the rat to electric shock, asphyxia, epinephrine, acetylcholine and serotonin but not to Pitressin, formalin, insulin, histamine and salicylate are completely blocked by midbrain transection.

A rise of plasma corticosterone levels in rats in response to the stress of ether anesthesia followed by withdrawal of blood sample by heart puncture [*Knigge and Hays,* 1963] and to etherization [*Kendall* et al., 1965] has been found to be significantly suppressed by midbrain lesions or transection. In contrast, *Davis* et al. [1961] failed to find any significant effect of midbrain transection on adrenal secretion of corticosterone and Porter-Silber chromogen in dogs subjected to laparotomy.

Evidence has accumulated that the effect of lesions or transection of the brain stem on pituitary-adrenocortical activity depends on the sites of lesions or the levels of transection.

Newman et al. [1958] studied the effect of midbrain transection or diencephalic-mesencephalic lesions on adrenocortical secretion in ether-chloralose-anesthetized cats subjected to adrenal vein cannulation. The adrenal cortisol secretion rate was not significantly altered by the transection of the mesencephalic reticular formation. It was, however, significantly reduced by lesions of the central core of the midbrain but not of the rostral pons. *Taylor and Farrell* [1962] showed that the adrenal cortisol secretion in cats anesthetized with ether-chloralose and subjected to adrenal vein cannulation was significantly elevated by placing large lesions in the rostral midbrain extending into the epithalamus and that it was somewhat reduced by discrete lesions limited to the rostral dorsal midbrain. *Slusher and Critchlow* [1959] reported that adrenal corticosterone secretion in rats in response to surgical stress (adrenal vein cannulation and 30-min collection of adrenal venous blood sample) was depressed by lesions in the rostral midbrain and posterior diencephalon, and enhanced by lesions in the posterior midbrain and rostral pons. They also reported that the adrenal ascorbic acid response in midbrain-lesioned rats to the above surgical stress was not correlated with the corticosterone secretory response. *Slusher* [1960] noted that adrenal corticosterone secretion in rats in response to surgical stress (laparotomy and adrenal vein cannulation) was found to be enhanced by the dorsal tegmental lesions and inhibited by the ventral tegmental lesions at the rostral pons level.

The effect of brain stem lesions on plasma corticosterone response in rats to traumatic stress (tibia fraction) has been extensively studied by *Gibbs* [1969a, b]. The rise of plasma corticosterone levels in response to unilateral tibia fraction (TF) was blocked by contralateral, but not ipsilateral, hemisection of the medulla oblongata. It was also found to be blocked by contralateral lesions and partially suppressed by ipsilateral lesions in the pontine reticular formation. Lesions in the pontine reticular formation were the most effective pontine lesions in suppressing the plasma corticosterone response to bilateral TF. The findings are interpreted to mean that the pituitary-adrenocortical secretory response in rats to traumatic stress is mediated by a nervous pathway which ascends to the pons mainly on the side contralateral to the traumatic area and that there exists a nervous structure in the pontine reticular formation which might facilitate the pituitary-adrenocortical secretory response to traumatic stress.

K. Spinal Cord Transection

Pituitary-adrenocortical secretory responses to some stressful stimuli, such as traumatic stress and sensory stimulation, have been shown to be blocked by spinal cord transection, when the stimuli are applied to a denervated area. Adrenal ascorbic acid depletion in rats in response to electrical stimulation of the hindpaws, but not forepaws, is abolished by spinal cord (T_1–T_3) transection [*Redgate*, 1960]. Adrenal corticosteroid secretion in the dog in response to traumatic stress is blocked by spinal cord transection [*Hume and Jackson*, 1959]. *Hume* et al. [1962] reported that adrenal 17-OHCS secretion in response to surgical stress (laparotomy) was abolished in a patient with spinal cord (T_4) transection.

However, pituitary-adrenocortical secretory responses to some other kinds of stressful stimuli have been found to be unaffected by spinal cord transection. A marked decrease in adrenal ascorbic acid concentration following ether anesthesia and unilateral adrenalectomy has been observed in rats with spinal cord (C_7) transection [*Martini* et al., 1960; *Giuliani* et al., 1961]. In conscious dogs with spinal cord (C_7 or C_{7-8}) transection, adrenal 17-OHCS secretion has been shown to be markedly increased after i.v. administration of *E. coli* endotoxin [*Egdahl*, 1959] and insulin [*Suzuki* et al., 1965c]. *Osborn* et al. [1962] failed to find any significant difference of peripheral plasma 17-OHCS response to surgical stress between patients with and without complete lesions of the spinal cord at the level above T_5.

Chapter 8. Neurotransmitter Mechanisms Involved in the Neural Control of Hypothalamic-Pituitary-Adrenocortical Activity

A. Catecholaminergic Mechanisms

From the finding that i.v. administration of *L*-dopa, which is known to cross the blood-brain barrier and increase the brain catecholamine content, significantly depresses the adrenocortical secretory response in pentobarbital-anesthetized dogs to surgical stress (laparotomy and intestinal manipulation), *Van Loon* et al. [1971a] have proposed a hypothesis that a catecholaminergic inhibition may be involved in the central nervous system mechanism for the control of pituitary-adrenocortical activity. Inhibition of the pituitary ACTH release by a central catecholaminergic mechanism has further been supported by the study of *Van Loon* et al. [1971b]. In their experiments the adrenal 17-OHCS secretion in pentobarbital-anesthetized dogs in response to surgical stress (laparotomy plus intestinal manipulation) was found to be significantly depressed or almost completely blocked by administration into the third ventricle of catecholamine (norepinephrine, dopamine or isoproterenol), which is ineffective when injected i.v., and by intraventricular injection of a systemically ineffective dose of catecholamine precursor (*L*-dopa), monoamine oxidase inhibitor (α-ethyltryptamine) or catecholamine releaser (tyramine).

In the above inhibitory central catecholaminergic mechanism in anesthetized dogs, dopamine might not be a neurotransmitter, since apomorphine (dopaminergic receptor stimulating drug) administered into the third ventricle of the brain is ineffective in depressing the stress-induced elevation of plasma corticosteroid levels and pimozide (dopaminergic receptor blocking drug) injected i.p. does not depress the inhibitory effect of *L*-dopa on stress-induced adrenocortical secretion [*Ganong* et al., 1976]. In pentobarbital-anesthetized dogs, intraventricular or i.v. administration of clonidine (α-adrenergic receptor stimulating drug) markedly inhibits the adrenocortical secretory response to surgical stress and intraventricular injection of phenoxybenzamine (α-adrenergic blocking drug) abolishes the inhibitory effect of L-dopa on adrenal corticosteroid secretion in response to surgical stress [Ganong et al., 1976]. Thus, it is suggested that an inhibitory neurotransmitter which participates in the control of pituitary-adrenocortical activity may be norepinephrine (or epinephrine).

The inhibitory effect of L-dopa on stress-induced adrenocortical secretion is abolished by i.v. injection of benserazide hydrochloride (dopa-decarboxylase inhibitor) which penetrates the blood-brain barrier [Ganong, 1977] but not by i.v. injection of carbidopa (dopa-decarboxylase inhibitor) which does not cross the blood-brain barrier [Ganong et al., 1976]. These findings suggest that the site of noradrenergic inhibition of pituitary-adrenocortical secretion might be located in the area inside the blood-brain barrier, most probably in the hypothalamic area dorsal to the median eminence. The concept of central noradrenergic inhibition of pituitary-adrenocortical activity was further supported by the study of Rose et al. [1976] who showed that the stress-induced elevation of plasma corticosteroid levels in pentobarbital-anesthetized dogs was depressed by electrical stimulation of the brain stem areas in or near the locus subceruleus or ventral ascending noradrenergic pathway but not by the stimulation of other brain stem areas.

In conscious cats, i.v. infusion of L-dopa has been found to produce a significant elevation of peripheral plasma 17-OHCS levels [Ruch et al., 1977]. In conscious dogs, Holland et al. [1978] have demonstrated that the peripheral plasma corticosteroid levels are significantly elevated by i.v. injection of L-dopa and this stimulatory effect of L-dopa is changed to depressing effect by pretreatment with pimozide (a dopamine receptor blocking drug, i.p.). In their experiments the peripheral plasma corticosteroid levels in conscious dogs has also been found to be significantly elevated by i.v. infusion of apomorphine (dopamine receptor stimulating drug) but not by i. v. injection of clonidine (α-adrenergic receptor stimulating drug). It is suggested that in conscious dogs a stimulatory central dopaminergic mechanism might participate in the regulation of pituitary-adrenocortical activity.

In the rat, evidence is rather controversial about the participation of catecholamines as neurotransmitters in the central nervous system regulation of pituitary-adrenocortical activity.

A significant elevation of plasma corticosterone levels has been observed by some investigators in rats whose hypothalamic catecholamines (norepinephrine and dopamine) are depleted by treatment with α-MpT (a tyrosine hydroxylase inhibitor, i.p. 3–9 h earlier) [*Scapagnini* et al., 1970, 1971b, 1975; *Van Loon* et al., 1971c]; a significant inverse correlation has been found in α-MpT-treated rats between plasma corticosterone levels and hypothalamic norepinephrine and dopamine concentrations. Depletion of hypothalamic catecholamines in rats induced by guanethidine (intraventricularly 8 h earlier) and depletion of hypothalamic norepinephrine induced by FLA-63 (a dopamine-β-hydroxylase inhibitor, i.p. 4 h earlier) have been found to be accompanied with a significant elevation of plasma corticosterone levels [*Scapagnini* et al., 1972]. *Cuello* et al. [1973/74] have shown that the α-MpT-induced elevation of plasma corticosterone levels in rats is partially depressed by *L*-threodihydroxyphenylserine (i.p.), which is known to be directly converted to norepinephrine. The above findings suggest that the central catecholaminergic mechanism is involved in the rat in neural control of the pituitary-adrenocortical secretory activity.

By contrast, *Carr and Moore* [1968] failed to find in rats at 4 h after i.p. injection of α-MpT, at which time a marked depletion in brain catecholamine was found, any significant changes in resting plasma corticosterone levels and plasma corticosterone responses to a variety of stressful stimuli, such as exposure to ether vapor, formalin (10% solution, s.c.), histamine (i.p.), restraint and transfer to a strange environment. *Bhattacharya and Marks* [1970] reported that in the rat the resting pituitary ACTH content, resting hypothalamic CRF activity, plasma and adrenal corticosterone responses to surgical stress (ether and sham laparotomy) or to chronic hypoxic stress (exposure to 10% O_2 + 90% N_2 for 16 h) and hypothalamic CRF response to the chronic hypoxic stress were not significantly altered at 16 h after administration of α-MpT (i.v.), at which time the brain norepinephrine in rats had been reported to be maximally depleted [*Brodie* et al., 1966]. In rats, brain norepinephrine depletion induced by i.p. administration of α-MpT every 3 h for 24 h is not accompanied with a significant elevation of plasma corticosterone levels [*McKinney* et al., 1971].

Pituitary-adrenocortical activity in rats whose hypothalamic catecholamines are depleted by intraventricular injection of 6-hydroxydopamine (6-OH dopamine), which is known to selectively destroy the catecholaminergic neurons, has been evaluated by a number of investigators. Studies on this line, however, revealed rather conflicting results.

Ulrich and Yuwiler [1973] failed to find any significant changes in circadian rhythms of plasma and adrenal corticosterone levels in rats whose whole brain norepinephrine and dopamine levels were reduced by pretreatment with 6-OH dopamine (intracisternally, twice 48 h apart 5–7 days earlier) to 28.7 and 53.4% of control values, respectively. *Kaplanski and Smelik* [1973a] observed that marked depletions of brain norepinephrine and dopamine in rats induced by 6-OH dopamine (intracisternally, twice with a 72-hour interval 7 days earlier or intraventricularly twice with a 48-hour interval 7 days earlier) were not accompanied by any significant changes in plasma corticosterone levels and in vitro corticosteroid production of excised adrenal glands. Similar findings were obtained by *Kaplanski* et al. [1973/74] in experiments with rats at 9–10 weeks of age, whose brain norepinephrine and dopamine were markedly depleted by 6-OH dopamine administered intracerebrally on days 1, 3 and 5 after birth. *Abe and Hiroshige* [1974] showed that a marked depletion of hypothalamic norepinephrine in rats induced by intraventricular administration of 6-OH dopamine (10–58 days earlier) did not affect the circadian rhythms of plasma corticosterone levels and of hypothalamic CRF activities; the findings suggest that hypothalamic norepinephrine is not essential in the rat for the regulation of pituitary-adrenocortical activity. *Honma and Hiroshige* [1979] also noted an essentially normal circadian rhythm of plasma corticosterone in 6-OH dopamine-treated rats. *Lippa* et al. [1973] reported that plasma corticosterone levels in rats were significantly depressed at 3 days and returned to the control levels at 11 days after intraventricular injection of 6-OH dopamine. They also noted that the plasma corticosterone response in rats to ketamine (i.p.) was significantly reduced 4 weeks after intraventricular injection of 6-OH dopamine. They stated that brain catecholamines in the rat might be stimulatory neurotransmitters in the neural pathway to the pituitary-adrenocortical system. In contrast, *Cuello* et al. [1974] demonstrated a significant decrease in hypothalamic norepinephrine accompanied by a significant elevation of plasma corticosterone levels in rats 12 or 24 h after 6-OH dopamine injections (i.p. twice with an interval of 12 h or intraventricularly twice with an interval of 48 h) and suggested an existence of inhibitory rather than stimulatory adrenergic mechanism involved in the central neural control of pituitary-adrenocortical secretory activity.

In the in vitro experiments with incubated rat hypothalamus, *Hill-house* et al. [1975] have demonstrated that acetylcholine-induced hypothalamic CRF release is suppressed by norepinephrine but not by dopamine and this inhibitory effect of norepinephrine is suppressed by phentolamine (α-adrenergic blocker). The experimental results suggest that norepinephrine may be an inhibitory neurotransmitter for the control of CRF release in the hypothalamus. In the in vitro experiments with synaptosomes prepared from sheep hypothalamic tissue, however, *Edwardson and Bennett* [1974] have found that acetylcholine-induced CRF release is markedly depressed by dopamine and only slightly by norepinephrine, suggesting that the inhibitory dopaminergic mechanism is involved in the regulation of hypothalamic CRF release.

B. Cholinergic Mechanisms

Evidence for the involvement of central cholinergic mechanism in the regulation of hypothalamic-pituitary-adrenocortical activity has been presented by *Endröczi* et al. [1963]. They have shown that a direct injection of carbaminoylcholine or eserine into the tuber cinereum, premammillary or posterior hypothalamic area, or into the ventral tegmentum or ventral reticular formation of the midbrain in conscious cats induces a moderate or marked elevation of adrenal corticosteroid secretion rate which is evaluated under pentobarbital anesthesia 45-60 min after the intracerebral injection. In conscious cats, *Krieger and Krieger* [1970b] have observed that implantation of carbaminoylcholine into the median eminence, central region of the mammillary body or amygdalar area results in an abrupt elevation of peripheral plasma levels of 11-OHCS. In rats, an injection of carbaminoylcholine into the lateral ventricle of the brain significantly elevates plasma corticosterone levels [*Abe and Hiroshige*, 1974].

However, s.c. injection of either of two anticholinesterases, galanthamine and neostigmine [*Naumenko*, 1967] and a direct injection of carbaminoylcholine into the posterior hypothalamic or rostral mesencephalic area [*Naumenko*, 1968] cause a significant elevation of plasma 17-OHCS levels in intact but not in midbrain-sectioned guinea pigs, suggesting that the stimulatory actions of these drugs may be mediated by some peripheral mechanisms. *Makara and Stark* [1976] reported that a significant elevation of plasma corticosterone levels was induced by the infusion of carbaminoylcholine into the third ventricle of the brain in intact rats, whereas it was not observed following the intraventricular injection of acetylcholine, carbaminoylcholine or oxotremorine (muscarinic stimulant) in rats with deafferented medial basal hypothalamus. The above studies failed to support the view that acetylcholine is a neurotransmitter in the medial basal hypothalamus which stimulates directly the CRF-releasing neurosecretory cells. However, a dose-dependent stimulatory effect of acetylcholine on the CRF release in vitro of the rat hypothalamic tissue [Bradbury et al., 1974; Hillhouse et al., 1975] but not of the rat median eminence tissue [Bradbury et al., 1974] has been demonstrated, suggesting that acetylcholine is a stimulatory neurotransmitter which may act at the dendritic levels of the CRF-releasing neurosecretory neurons.

Implantation of atropine in the anterior hypothalamus in pentobarbital-anesthetized rats significantly depresses an increase in pituitary-adrenocortical activity (corticosteroid production in vitro of excised adrenal glands) induced by surgical stress of the hypothalamic implantation procedure or exposure to ether vapor [Hedge and Smelik, 1968; Kaplanski and Smelik, 1973b], i.v. injection of arginine vasopressin [Hedge and Smelik, 1968], i.m. injection of carbaminoylcholine or epinephrine, i.v. injection of histamine, laparotomy plus intestinal traction, and unilateral adrenalectomy [Kaplanski and Smelik, 1973b]. It has also been noted that hypothalamic implantation of atropine depresses the elevation of plasma corticosterone levels in rats in response to surgical stress of the implantation procedure [Hedge and Smelik, 1968; Hedge and De Wied, 1971; Kaplanski and Smelik, 1973b]. These findings may be interpreted to indicate that the cholinergic mechanism in the hypothalamus is involved in the stress-induced elevation of hypothalamic-pituitary-adrenocortical activity. However, Makara and Stark [1976] have suggested that the inhibitory effect of atropine implants in the anterior hypothalamus in rats does not necessarily indicate the participation in stress-induced elevation of pituitary-adrenocortical activity of cholinergic synapses in the anterior hypothalamus. In conscious cats, cerebrospinal perfusion with atropine does not alter the peripheral plasma cortisol levels, showing that acetylcholine is not an essential neurotransmitter for maintaining the basal activity of the pituitary-adrenocortical system [Garcy and Marotta, 1978].

C. Serotoninergic Mechanisms

Naumenko [1968] has shown that plasma 17-OHCS levels in guinea pigs are significantly elevated by the direct injection of serotonin into the medial parts of the posterior, middle or anterior hypothalamic area or into the preoptic area but not into the lateral hypothalamus, and this plasma 17-OHCS response is not abolished by the midbrain section. The above findings may suggest the possibility that serotonin is a stimulatory transmitter between the afferent terminal neurons in the hypothalamus and the CRF-releasing neurosecretory neurons. *Krieger and Krieger* [1970b] have observed that in unrestrained conscious cats an implantation of serotonin in the median eminence or septum but not in the mammillary body, hippocampus, and lateral and basal amygdalar areas results in an elevation of peripheral plasma 11-OHCS levels. *Popova* et al. [1972] have demonstrated that in intact rats and in rats with complete deafferentation of the medial basal hypothalamus an elevation of hypothalamic serotonin levels induced by systemic administration of 5-hydroxytryptophan (5-HTP, a serotonin precursor) is accompanied by a marked increase in plasma corticosterone concentration; the possibility of direct action of 5-HTP at the adrenal level has been excluded in their study by the finding that plasma corticosterone response in rats to 5-HTP is completely abolished by hypophysectomy. They suggested that serotonin might participate as a stimulatory transmitter in the neural control of hypothalamic CRF release.

The concept of stimulatory serotoninergic mechanism in the neural control of pituitary-adrenocortical activity has been supported by *Fuller* et al. [1976]. They have found that plasma corticosterone levels in rats are markedly elevated by i.p. administration of fluoxetine (3-(*p*-trifluoromethylphenoxy)-3-phenyl-*N*-methyl-propylamine, Lilly 110140), which is known to prevent serotonin reuptake, and of 5-HTP, a precursor of serotonin. They have also found that the plasma corticosterone response in rats to fluoxetine is abolished by hypophysectomy. It has also been noted that the plasma corticosterone response in rats to 5-HTP is enhanced by pretreatment with fluoxetine [*Fuller* et al., 1975] and that plasma corticosterone levels in rats are markedly elevated by simultaneous administration of low doses of 5-HTP and fluoxetine, in which dose neither agent is effective when given alone [*Fuller* et al., 1976]. *Marotta* et al. [1976] have also noted a significant elevation of plasma corticosterone levels in rats in response to fluoxetine.

By contrast, evidence for the inhibitory action of serotonin on the pituitary-adrenocortical system has also accumulated. *Vermes and Tel-egdy* [1972] have reported that the plasma corticosterone response in rats to ether stress is significantly impaired by implantation of serotonin into the medial hypothalamus but not into the anterior or posterior hypothalamic area, whereas the basal plasma corticosterone levels are unaltered by serotonin implantation into the medial, anterior or posterior hypothalamus. Inhibitory action in vitro of serotonin on hypothalamic-pituitary-adrenocortical activity has been shown by *Vermes* et al. [1972]. The rise of plasma corticosterone levels in rats following exposure to ether vapor is significantly inhibited by pretreatment with 5-HTP [*Vernikos-Danellis* et al., 1977].

However, *Dixit and Buckley* [1969] have shown that the adrenal ascorbic acid depletion and the elevation of plasma corticosterone in response to cold stress or shaker stress are not altered in rats whose brain serotonin is markedly depleted by pretreatment with 4-chloramphetamine (i.p.) or by feeding a tryptophan-deficient diet and in rats whose brain serotonin levels are significantly elevated by pretreatment with tranylcypromine (a monoamine oxidase inhibitor, i.p.). Depletion of brain serotonin in rats by pretreatment with p-chlorophenylalanine (p-CPA, an inhibitor of serotonin biosynthesis, i.p.) does not significantly affect the plasma and adrenal corticosterone responses to surgical stress (ether and sham laparotomy) [*Bhattacharya and Marks,* 1970] and plasma corticosterone levels following a 1-hour exposure to hypoxic stress (10% O_2 + 90% N_2) and hypercapnic stress (10% CO_2 + 20% O_2 + 70% N_2) [*Marotta* et al., 1976]. In cats, the plasma 17-OHCS response to either insulin hypoglycemia or bacterial polysaccharide of *Pseudomonus* origin has been found to be unaffected by pretreatment with drugs which alter brain serotonin levels or serotonin action [*Krieger and Rizzo,* 1969].

As to the effect of brain serotonin depletion on basal plasma corticosteroid levels, *Preziosi* et al. [1968] failed to find any significant change in corticosterone levels in rats 2, 6, 12, 24 and 48 h, and 3–7 days after administration of p-CPA (i.p.). In contrast, *Marotta* et al. [1976] observed a significant elevation of plasma corticosterone levels in rats 48 h after injection of p-CPA (i.p.).

On the other hand, *Scapagnini* et al. [1971a], *Van Delft* et al. [1973] and *Vernikos-Danellis* et al. [1973] reported that depletion of brain serotonin in rats induced by pretreatment with p-CPA (i.p.) resulted in an elevation of morning levels and a depression of evening levels, thus a partial or complete abolition of an overt circadian rhythm, of plasma corticosterone. *Szafarczyk* et al. [1979] observed significant elevations of daily minimum and mean levels, and a significant decrease in the amplitudes of circadian variation, of plasma corticosterone in p-CPA-treated rats. *Krieger and Rizzo* [1969] have also noted that the circadian rhythms of peripheral plasma 17-OHCS levels in cats are abolished by treatment with drugs which alter brain serotonin levels or serotonin action. The above findings may be interpreted to indicate that the serotoninergic mechanism in the brain is essential to the circadian periodicity of pituitary-adrenocortical activity. However, *Rotsztejn* et al. [1977] showed that the circadian periodicity of plasma corticosterone levels was not abolished in rats whose serotonin levels in the hypothalamus and dorsal hippocampus were depressed by treatment with p-CPA (i.p.) or by lesions of the midbrain raphe nuclei.

D. Gabanergic Mechanisms

In unrestrained conscious cats, implantation of γ-aminobutyric acid (GABA) into the median eminence or lateral amygdalar areas has been found to elevate peripheral plasma 11-OHCS levels [*Krieger and Krieger,* 1970b]. In rats, however, *Abe and Hiroshige* [1974] failed to find any significant change in plasma corticosterone levels following injection of GABA into the lateral ventricle of the brain.

Makara and Stark [1974] have observed that the plasma corticosterone response in rats to surgical stress (unilateral sham adrenalectomy) is completely prevented by infusion of GABA into the third ventricle of the brain. They have also observed that plasma corticosterone levels in rats with deafferented medial basal hypothalamus are significantly elevated by the infusion into the third ventricle of either of two GABA antagonists, picrotoxin and bicuculline; the plasma corticosterone response to the latter GABA antagonist is strongly depressed by simultaneous infusion of GABA. The above findings suggest that GABA may be involved in the regulatory mechanism of hypothalamic-pituitary-adrenal activity as an inhibitory rather than a stimulatory neurotransmitter of the hypothalamic interneurons or afferent pathways.

Chapter 9. Feedback Regulation of Pituitary-Adrenocortical Activity

A. Corticosteroid Feedback Inhibition of Pituitary-Adrenocortical Activity

Feedback inhibition of pituitary-adrenocortical activity by corticosteroids has been suggested by numerous observations, such as an adrenal atrophy following administration of large doses of adrenocortical extract in normal but not in corticotropin-treated hypophysectomized animals, a compensatory adrenal hypertrophy following unilateral adrenalectomy, a marked depletion of pituitary ACTH after long-term treatment with corticosteroids, an increase in pituitary ACTH content following bilateral adrenalectomy and a marked depletion of adrenal corticosteroids following treatment with corticosteroids.

It has been noted that stress-induced depletion of adrenal ascorbic acid in rats is suppressed or completely blocked by pretreatment with corticosteroids [*Sayers and Sayers*, 1947; *Gray and Munson*, 1951; *Abelson and Baron*, 1952; *Ohler and Sevy*, 1956; *Wexler* et al., 1958; *De Wied and Mirsky*, 1959; *Ashford and Shapero*, 1962; *Vernikos-Danellis and Marks*, 1962; *Hodges and Sadow*, 1967]. Suppression by corticosteroids of adrenal ascorbic acid depletion in cats has also been reported [*Schwartz and Kling*, 1960].

Adrenal 17-OHCS (or cortisol) or 11-OHCS secretion in response to stressful stimuli has been shown to be suppressed by treatment with corticosteroids in pentobarbital-anesthetized dogs [*Richards and Pruitt*, 1957; *Myers* et al., 1960; *Egdahl*, 1964; *Asano*, 1966; *Zukoski*, 1966; *Nakasone and Shimizu*, 1971; *Marotta* et al., 1973] and in conscious dogs [*L'Age* et al., 1970].

Plasma corticosterone levels are significantly depressed in nonstressed corticosteroid-treated rats [*Pfeiffer* et al., 1960; *Sen and Sa-*

rangi, 1967; *Zimmermann and Critchlow,* 1969a, b; *Zimmermann* et al., 1972; *Buckingham and Hodges,* 1976]. Resting plasma corticosteroid levels in newborn dogs [*Muelheims* et al., 1969], in cats [*Rivas and Borrell,* 1971] and in rhesus monkeys [*Setchell* et al., 1975a] are also depressed by treatment with corticosteroids.

It has been reported by a vast number of investigators that a stress-induced elevation of plasma corticosterone levels in rats is suppressed by pretreatment with corticosteroids [*Yates* et al., 1961; *Leeman* et al., 1962; *Mangili* et al., 1965; *Hodges and Sadow,* 1967; *Sirett and Gibbs,* 1969; *Moberg,* 1971; *Takebe* et al., 1971a; *Jones* et al., 1972; *Kendall* et al., 1972; *Riegle and Hess,* 1972; *Dallman and Jones,* 1973; *Feldman* et al., 1973; *Kuwamura* et al., 1973; *Jones* et al., 1974; *Feldman and Conforti,* 1976a; *Jobin* et al., 1976; *Abe and Critchlow,* 1977; *Tang and Phillips,* 1977; *Kaneko and Hiroshige,* 1978a; *Sithichoke and Marotta,* 1978; *Vernikos* et al., 1982]. Stress-induced elevation of plasma 11-OHCS in conscious cats is blocked by prior administration of corticosteroids [*Krieger and Krieger,* 1970a].

Sayers and Sayers [1947] have studied the quantitative relationship between the stimulus intensity and the amount of corticosteroids required to suppress the pituitary-adrenocortical response. They have observed that the adrenal ascorbic acid response in rats is suppressed by treatment with adequate amount of corticosteroids and that the amount of corticosteroids required to suppress the response is proportional to the intensity of the ACTH-releasing stimulus. They proposed that stressful stimuli would increase the utilization of corticosteroids by peripheral tissues and depress the circulating corticosteroid levels; this depression of corticosteroid levels would in turn reduce the corticosteroid suppression of the pituitary activity and thus stimulate the ACTH release.

Subsequent studies by a number of investigators, however, have shown an almost immediate elevation rather than a fall in plasma corticosteroid levels and prolongation of the half-life of corticosteroids following exposure to stressful stimuli. It has also been reported that an increase in plasma ACTH concentration can be induced in bilaterally adrenalectomized animals. Thus, the hypothesis of *Sayers and Sayers* [1947] has been disproved.

Later, a variable set-point control hypothesis was proposed by *Yates* et al. [1961]. They evaluated in pentobarbital-anesthetized rats the effect of i.v. injection of histamine or laparotomy with or without preceding administration of corticosterone on plasma corticosterone levels. In histamine experiments, corticosterone was administered in a dose sufficient to produce elevation of plasma corticosterone levels equal to that observed after injection of histamine alone. In rats injected with histamine 15–30 s after administration of corticosterone, the rise of plasma corticosterone levels was not significantly different from that in rats administered with corticosterone alone. In laparotomy experiments, a small (5 or 12 μg/100 g) or large (35 μg/100 g) dose of corticosterone was administered before laparotomy. In rats injected with a small dose of corticosterone 15–30 s before laparotomy, a rise in plasma corticosterone levels was the same as that in rats subjected to laparotomy without preceding corticosterone injection. In rats injected with a large dose of corticosterone just before laparotomy a rise in plasma corticosterone levels was exactly the same as that in rats injected with corticosterone alone. From these findings they proposed the hypothesis of the negative feedback controller with a variable set-point which participates in the regulation of the pituitary ACTH release. They stated that a rapid resetting of the controller to a high set-point level induced by stressful stimuli resulted in a virtual but not an actual fall in plasma corticosterone levels, which in turn stimulated ACTH release of the pituitary. However, failure to agree with the above variable set-point control hypothesis has been reported by Smelik [1963].

Dallman and Yates [1969] have provided evidence for the existence of the fast and delayed feedback inhibitions by corticosteroids of the stress-induced pituitary ACTH release. They have examined plasma corticosterone response in pentobarbital-anesthetized rats to histamine (230 μg, i.v.) injected at various times after the onset of i.v. infusion of corticosterone at the rate of 1.8–1.9 μg/min and have observed corticosteroid feedback inhibition soon after the onset of the infusion when plasma corticosterone levels are rising rapidly. The period of the fast, rate-sensitive feedback, which is of short duration (a few min), is followed by a period during which no feedback inhibition is observed (a silent period).

Subsequently, characteristics of the fast, rate-sensitive feedback inhibition were studied by Jones et al. [1972]. They evaluated adrenal and plasma corticosterone responses in pentobarbital-anesthetized rats to surgical stress (sham adrenalectomy) which was applied at various times after the onset of i.p. infusion of corticosterone (8 μg/min). Complete suppression of the pituitary-adrenocortical responses to surgical stress was observed whenever plasma corticosterone levels were rising at a rate exceeding 1.3 μg/100 ml/min. Feedback inhibition was found to be abolished when the total dose of corticosterone was raised either by increasing the duration of corticosterone infusion or by increasing corticosterone concentration in the infusate. Thus, the phenomenon of 'fast feedback saturation' was clearly demonstrated.

Jones et al. [1974] showed two distinct periods of feedback inhibition, fast and delayed feedback, of plasma corticosterone response to ether stress (exposure to ether vapor for 2 min) in rats following administration of corticosterone (0.1–5 mg/100 g, s.c.). Small doses of corticosterone were found to be effective in inducing fast feedback inhibition which was observed within a few minutes after administration of corticosterone. The duration of the fast feedback inhibition was short and it was further shortened by administration of larger doses of corticosterone. Delayed feedback inhibition occurred after 1 h or later and required larger doses of corticosterone. Duration of delayed feedback inhibition was increased by increasing doses of corticosterone. Saturation of fast feedback and summation of delayed feedback were observed when a combination of inhibitory doses of corticosterone and cortisol was given.

Abe and Critchlow [1977] evaluated the plasma corticosterone response in pentobarbital-anesthetized rats to ether stress (exposure to ether vapor for 2 min) which was applied at various times after the onset of 10-min i.p. infusion of corticosterone (4 or 8 μg/min). The response was found to be blocked when plasma corticosterone levels were rising at the rate of 2.1–4.6 μg/100 ml/min but not of 0.7 μg/100 ml/min. Thus, in the fast, rate-sensitive feedback inhibition the minimum effective rate of rise in plasma corticosterone levels for suppressing plasma corticosterone response to ether stress would be between 0.7 and 2.1 μg/100 ml/min.

Kaneko and Hiroshige [1978a] performed quantitative analyses of fast, rate-sensitive feedback inhibition of stress-induced pituitary ACTH release. They determined the plasma corticosterone response in pentobarbital-anesthetized rats to histamine (0.23, 0.46 and 1.15 mg, i.v.) injected at various times after the onset of i.v. infusion of corticosterone. The minimum effective rate of rise in plasma corticosterone levels for inhibiting histamine-induced plasma corticosterone elevation was found to be 4–6 μg/100 ml/min.

B. Sites and Mechanisms of Negative Feedback Action of Corticosteroids

In vitro Studies with Pituitary Fragments or Isolated Pituitary Cells

Suppression by corticosteroids of ACTH release of isolated pituitary cells or pituitary fragments in vitro has been studied by a number of investigators.

ACTH release from incubated rat anterior pituitary fragments induced by vasopressin [*Fleischer and Vale,* 1968], dibutyryl cyclic AMP [*Fleischer* et al., 1969], high concentration of K^+ [*Kraicer* et al., 1969] and rat SME (or hypothalamic) extract [*Arimura* et al., 1969; *Berthold* et al., 1970; *Kraicer and Milligan,* 1970; *Buckingham and Hodges,* 1977] is inhibited by addition of corticosteroids into the incubation medium. ACTH release of isolated incubated rat anterior pituitary cells [*Portanova and Sayers,* 1973b, 1974] and of isolated superfused rat anterior pituitary cells [*Mulder and Smelik,* 1977; *Gillies and Lowry,* 1978] induced by rat hypothalamic-median eminence extract is suppressed by corticosterone. Thus, direct inhibition of stimulus-induced ACTH release by corticosteroids at the pituitary level in vitro has been clearly demonstrated. As to the direct effect of corticosteroids on spontaneous ACTH release of the pituitary in vitro, however, data appear to be in conflict. Spontaneous release of ACTH has been found to be suppressed by cortisol in in vitro experiments with bovine anterior pituitary fragments [*Pollock and LaBella,* 1966], by corticosterone in in vitro experiments with rat anterior pituitary fragments [*Kraicer and Milligan,* 1970], and by corticosterone and dexamethasone in a monolayer culture of rat pituitary tissue [*Fleischer and Rawls,* 1970]. In contrast, it has been reported that spontaneous ACTH release is not suppressed by dexamethasone in rat pituitary tissue culture [*Stark* et al., 1968] and in rat anterior pituitary fragment incubation [*Fleischer and Vale,* 1968], and by corticosterone in in vitro incubation of rat anterior pituitary fragments [*Buckingham and Hodges,* 1977] and in isolated rat anterior pituitary cell superfusion [*Mulder and Smelik,* 1977].

Intrahypothalamic and Intrapituitary Implantation (or Injection) of Corticosteroids

Rose and Nelson [1956] reported that cortisol, continuously infused into the hypophyseal fossa for 8 days in unilaterally adrenalectomized rats, produced a significantly greater suppression of compensatory adrenal hypertrophy than the same dose of cortisol infused s.c. or lateral to the hypophyseal fossa. The finding suggests that corticosteroids act directly at the pituitary level to suppress ACTH release. However, *Kendall* [1962] observed that direct microinjection of various doses of dexamethasone (1–40 µg/100 g) into the rat pituitary was no more effective in inhibiting adrenal corticosterone secretion than s.c. injection of the same doses of dexamethasone, suggesting that the receptor area for corticosteroid feedback inhibition is not localized in the pituitary. In pentobarbital-anesthetized dogs subjected to surgical stress (adrenal vein cannulation), *Stark* et al. [1968] failed to find any suppressing effect of intrapituitary infusion of dexamethasone on adrenal secretions of cortisol and corticosterone. They stated that corticosteroids did not suppress ACTH release at the pituitary level under certain stressful conditions.

In rats, implantation of small amounts of corticosteroids into the median eminence but not into the pituitary has been found to prevent compensatory adrenal hypertrophy following unilateral adrenalectomy [*Davidson and Feldman,* 1963; *Feldman* et al., 1965/66], to produce adrenal atrophy and depress surgical stress-induced adrenal ascorbic acid depletion [*Chowers* et al., 1963], and to depress adrenal and plasma corticosterone levels, and adrenal weight [*Corbin* et al., 1965]. The plasma corticosterone response in rabbits to immobilization stress has been observed to be inhibited by corticosteroid implantation into the anterior part of the median eminence but not into the anterior pituitary [*Smelik and Sawyer,* 1962].

The above experimental results in rats and in rabbits appear to show that the site of feedback action of corticosteroids may be located in the median eminence but not in the pituitary.

By contrast, it has been reported that implantation or microinjection of corticosteroids in the anterior pituitary as well as in the median eminence depresses adrenal weight and pituitary ACTH content [*Chowers* et al., 1967], plasma corticosterone response to scald [*Russell* et al., 1969], and ACTH release (corticosteroid production of excised adrenal glands in vitro) in response to ether stress [*Bohus and Strashimirov,* 1970]. These findings may be interpreted to indicate that both the median eminence and the anterior pituitary are the sites of corticosteroid feedback action.

Implantation of corticosteroids in rat median eminence has been noted to depress adrenal weight, adrenal corticosteroid content and p.m. plasma corticosteroid levels [*Slusher*, 1966], plasma corticosterone response to burn [*Dallman and Yates*, 1967], adrenal weight [*Feldman* et al., 1967], plasma and adrenal corticosterone levels and adrenal corticosterone secretion [*Kendall and Allen*, 1968], adrenal and pituitary weights, basal a.m. plasma corticosterone levels and plasma corticosterone responses to ether stress (etherization and venesection) and electric shock [*Davidson* et al., 1968], adrenal weight and plasma corticosterone response to ether stress [*Grimm and Kendall*, 1968], and stress (transfer of the animal to a new environment)-induced and basal release of ACTH (corticosteroid production of excised adrenal glands in vitro) [*Bohus* et al., 1968].

It has been reported that the pituitary-adrenocortical activity in rats is suppressed by corticosteroid implants in various hypothalamic areas; e.g. the caudal hypothalamus [*Endröczi* et al., 1961], basomedial hypothalamus [*Bohus and Endröczi*, 1964b], anterior hypothalamus [*Davidson and Feldman*, 1967], and anterior and posterior hypothalamus [*Zimmermann and Critchlow*, 1969c].

Collectively, the inhibitory effect of corticosteroid implants in the median eminence or some other hypothalamic areas on pituitary-adrenocortical activity has been well established, whereas studies on the effect of intrapituitary corticosteroid implants have revealed rather conflicting results. Thus, for instance, *Davidson and Feldman* [1963] stated that the hypothalamus rather than the pituitary should be the primary locus of corticosteroid feedback inhibition, whereas *Bohus and Strashimirov* [1970] postulated that the receptors for corticosteroid feedback inhibition may be located both in the hypothalamus and anterior pituitary. However, a localized and direct action of implanted corticosteroids at the hypothalamic level cannot be expected in hypothalamic implantation experiments, since there is a possibility, which cannot be excluded, that a part of implanted corticosteroids may spread from the median eminence via the hypophyseal portal circulation toward the anterior pituitary and may exert an inhibitory action at the pituitary level. Indeed, *Bogdanove* [1963] demonstrated in the studies of the effect of intrahypothalamic and intrapituitary estrogen implantations in rats on gonadotropin secretion that estrogen could be distributed to the entire anterior pituitary by the intrahypothalamic but not intrapituitary implantation; estrogen could be supplied by the former implantation to the median eminence and then it might be distributed via hypophyseal portal circulation to the entire anterior pituitary, whereas it could be distributed by the latter implantation only to adjacent regions in the anterior pituitary. He postulated the concept of 'implantation paradox' – i.e. the most efficient method for eliciting the direct effect of steroids at the pituitary level in vivo might be the steroid implantation into the median eminence or into some other hypothalamic areas.

The implantation paradox may be true of the feedback inhibition of ACTH release by implanted corticosteroids. *Kendall and Allen* [1968] studied the sites of corticosteroid feedback inhibition on ACTH release in hypophysectomized rats bearing 10 heterotopic hypophyses transplanted under the renal capsules. They observed that adrenal secretion of corticosterone was significantly depressed by i.p. injection of a large dose of dexamethasone in hypophysectomized, pituitary-transplanted (P-T) rats as well as in intact rats. Implantation of cortisol into the median eminence in hypophysectomized P-T rats was found to be ineffective in depressing adrenal corticosterone secretion, and adrenal and plasma corticosterone levels, whereas it was effective in intact rats. Adrenal atrophy was observed 2 weeks after the median eminence implantation of cortisol in intact but not hypophysectomized P-T rats. The findings suggest that in hypophysectomized P-T rats the median eminence is not essential in feedback control by corticosteroids and the site of inhibitory action of corticosteroids is located in the transplanted pituitaries. They also stated that in intact rats the median eminence might be important for feedback inhibition by corticosteroids only because of its location adjacent to the primary plexus of the hypophyseal portal veins and an actual site of inhibition might be located in the anterior pituitary rather than the median eminence. *Russell* et al. [1969] suggested that the ineffectiveness of intrapituitary corticosteroid implants in suppressing ACTH release might probably be due to an inadequate exposure of the anterior pituitary tissue to corticosteroids.

Implantation (or Microinjection) of Corticosteroids into the Extrahypothalamic Brain Structures

The effect of corticosteroid implantation or microinjection into a variety of extrahypothalamic brain structures in the rat on pituitary-adrenocortical activity has been studied by a number of investigators. Adrenal corticosteroid secretion is suppressed by implantation of cortisol or cortisone into the mesencephalic reticular formation but not into the thalamus or the preoptic area or cerebral cortex [*Endröczi* et al., 1961; *Bohus and Endröczi*, 1964b]. Adrenal and plasma corticosterone levels are depressed by cortisol implantation into the lateral mesencephalic reticular formation but not into the mesencephalic periventricular gray or cerebral cortex [*Corbin* et al., 1965]. Resting p.m. levels of adrenal and plasma corticosteroids are depressed by implantation of cortisol in the mesencephalic reticular formation or ventral hippocampus [*Slusher*, 1966]. Compensatory adrenal hypertrophy is completely prevented by dexamethasone implantation into the various forebrain areas extending from the preoptic area to the septum and adjacent areas [*Davidson and Feldman*, 1967]. The plasma corticosterone response to burn is suppressed by injection of dexamethasone into the anterior thalamus or septal areas [*Dallman and Yates*, 1967]. Pituitary ACTH release (corticosteroid production of excised adrenal glands in vitro) in response to stress of transfer to a new environment, but not the basal ACTH release, is suppressed by implantation of cortisol in the amygdala, basal septum, medial thalamus and the rostral mesencephalic reticular formation [*Bohus* et al., 1968]. Basal plasma corticosterone levels but not the plasma corticosterone response to ether stress are depressed by implantation of dexamethasone into the midbrain tegmentum or thalamus but not into the amygdala [*Zimmermann and Critchlow*, 1969c].

In rabbits, corticosterone implantation into the central, lateral basal and medial nuclei of the amygdala produces a decrease in pituitary-adrenocortical activity (incorporation of [1-^{14}C]acetate into adrenal corticosteroids in vivo) [*Kawakami* et al., 1968b]. In pentobarbital-anesthetized cats, microinjection of cortisone into the mesencephalic reticular formation results in a depression of adrenal corticosteroid secretion [*Endröczi* et al., 1961].

The inhibitory effect of corticosteroid implants in the extrahypothalamic forebrain structures on pituitary ACTH release does not necessarily indicate the existence of the site of inhibitory action in the loci of corticosteroid implantation, since a possibility of corticosteroid spread from the sites of implantation to the median eminence or anterior pituitary is not excluded. In fact, the sites of corticosteroid implantation in the forebrain structures noted in the above studies are located close to the ventricular system. *Kendall* et al. [1969] reported that in rats implants of dexamethasone in the lateral ventricle and implants of corticosterone in the third ventricle or brain structures immediately adjacent to the third ventricle were found to suppress plasma corticosterone response to ether stress. They stated that the inhibitory effect of corticosteroids implanted in the extrahypothalamic brain structures in rats could be explained on the basis of corticosteroid transport from the sites of implantation through the ventricular system to a site of feedback inhibition located in the basal hypothalamus or anterior pituitary. *Russell* et al. [1969] have shown that the plasma corticosterone response to i.v. injected CRF (prepared from ovine SME tissue), which is known to stimulate directly the anterior pituitary cells, is suppressed by microinjection of dexamethasone into the area adjacent to the septum of the forebrain. The findings are interpreted to show the spread of dexamethasone to the anterior pituitary. The same investigators have also demonstrated in rats the spread of eosin to the anterior pituitary within 5 min after the septal microinjection. Passage of corticosteroids from the cerebrospinal fluid in the third ventricle to hypophyseal portal vein blood has been clearly shown by *Ondo* et al. [1972]; within 30 min after injection of [4-^{14}C]corticosterone into the third ventricle of the rat, a significant quantity of radioactive corticosterone was recovered in the blood of the hypophyseal portal vein.

Studies in Animals with Hypothalamic Deafferentation or Midbrain Section

The inhibitory effect of dexamethasone (i.p.) on plasma corticosterone response in rats to ether stress has been found to be reduced to some extent by complete deafferentation of the medial basal hypothalamus [*Feldman* et al., 1973] and posterior hypothalamic deafferentation [*Feldman and Conforti*, 1976a], suggesting a participation of brain structures outside the medial basal hypothalamus in the feedback control of the pituitary-adrenocortical system by circulating corticosteroids. In midbrain-sectioned rats a significant depression in plasma corticosteroid levels is induced by dexamethasone in a dose of 100 μg/100 g, but not 25–50 μg/100 g, whereas it is induced in intact rats by 3 μg/100 g dexamethasone. Thus, the responsiveness to corticosteroids in feedback inhibition has been found to be markedly reduced in midbrain-sectioned rats [*Fraschini* et al., 1964]. The findings suggest a participation of the midbrain as a modulator in the feedback control of pituitary-adrenocortical activity.

Evaluation of Pituitary-Adrenocortical Responses to
Exogenous CRF

Suppression by pretreatment with corticosteroids of exogenous CRF-induced elevation of pituitary-adrenocortical activity has been observed by some investigators, a finding which may be interpreted to mean that at least one of the sites of inhibitory feedback action of corticosteroids is located in the anterior pituitary.

Adrenal ascorbic acid depletion in intact rats induced by i.v. injection of CRF (calf SME extract) has been found to be abolished by pretreatment with deoxycorticosterone acetate [*Royce and Sayers,* 1959]. In rats with basal hypothalamic lesions the elevation of pituitary-adrenocortical activity, as judged by an increase in corticosteroid production in vitro of excised adrenal glands, following i.v. injection of CRF (rat hypothalamic extract) is suppressed by pretreatment with dexamethasone [*De Wied,* 1964] or with corticosterone [*Jones* et al., 1977]. In conscious dogs, an increase in adrenal secretion of cortisol induced by intrapituitary injection of CRF (extract of ovine SME tissue) has been shown to be completely suppressed by pretreatment with dexamethasone (s.c.) [*Gonzalez-Luque* et al., 1970].

In contrast, *Leeman* et al. [1962] found that the plasma corticosterone response in rats to i.v. injected CRF (calf hypothalamic extract) was not affected by physiological doses of corticosterone (i.v.) which were capable of blocking the plasma corticosterone response to histamine (i.v.). The findings are interpreted to show that the sites of inhibitory corticosteroid action on stress-induced elevation of plasma corticosterone levels might be located in the brain but not in the anterior pituitary. Similar observation was also reported by *Hedge and Smelik* [1969]. They observed that the elevation of pituitary-adrenocortical activity (corticosteroid production of excised adrenal glands in vitro) in rats induced by i.v. injected CRF (prepared from the calf brain stem extract) was not suppressed by pretreatment with 25 µg/rat of dexamethasone injected s.c. 2 h previously, a dose sufficient to completely block the pituitary-adrenocortical response to ether or surgical stress.

Evaluation of the Hypothalamic CRF Activity

If non-stress circulating corticosteroids exert a suppressive action on ACTH release by inhibiting directly or indirectly the hypothalamic CRF release or synthesis, bilateral adrenalectomy would result in a change in hypothalamic CRF activity. However, experimental results obtained in the studies on this line have been controversial.

   Vernikos-Danellis [1965] evaluated the CRF activity of the rat median eminence at various time intervals after adrenalectomy. It was found that the CRF activity decreased at 1 and 3 h, returned to the control level by 6 h and began to increase 5 days after adrenalectomy. It peaked by 20 days at which time it was 2 or 2.5 times that of control value and remained high for another 30 days. An increase in hypothalamic CRF activity in rats 4 days after adrenalectomy was also reported by *Buckingham* [1979].

   In contrast, *Hiroshige* et al. [1969b] failed to find any significant change in CRF activity of the median eminence in rats on the 11th and 20th days after adrenalectomy. It was also noted that there was no significant change in rat hypothalamic CRF activity 15 days [*Yasuda and Greer*, 1976b], 1 day and 14 days [*Hillhouse and Jones*, 1976] after adrenalectomy, although a significant decrease in hypothalamic CRF activity was found by the latter investigators at 2 h after adrenalectomy. A decrease by 50% in rat hypothalamic CRF activity was observed by *Seiden and Brodish* [1971] 3–6 weeks after adrenalectomy, although there was no significant change at 1 week.

   The effect of corticosteroids in vivo on hypothalamic CRF activity has also been examined by a number of investigators. *Vernikos-Danellis* [1965] reported that a rise in CRF activity of the median eminence in rats induced by ether stress plus surgical stress (sham unilateral adrenalectomy) was completely abolished and non-stress CRF activity was markedly reduced by cortisol given s.c. 4 h before the observation. She suggested that cortisol exerted a negative feedback action primarily by suppressing the synthesis of hypothalamic CRF. However, *Hiroshige* et al. [1969b] and *Yasuda and Greer* [1976b] failed to find any significant change in non-stress hypothalamic CRF activity following treatment with dexamethasone, although the former investigators observed that a rise in median eminence CRF activity in rats induced by exposure to ether vapor followed by laparotomy was depressed by dexamethasone given i.p. 4 h previously. A rise in median eminence CRF activity in rats in response to laparotomy plus intestinal traction has been shown to be suppressed by pretreatment with dexamethasone (i.p.), suggesting that the site of feedback inhibition by corticosteroids may be located mainly in the hypothalamus or extrahypothalamic brain structures [*Takebe* et al., 1971a].

In the studies of *Sakakura* et al. [1976a], rats were subjected to a 2-min immobilization stress twice at a time interval of 5 or 23 min. When the second immobilization was started after 5 min, at which time plasma corticosterone levels were rising, a rise in hypothalamic CRF activity in response to the second immobilization was found to be blocked. However, it was not blocked when the second immobilization was performed at 23 min, at which time plasma corticosterone attained to and remained at high plateau levels. They stated that the hypothalamic CRF activity is influenced by the fast, rate-sensitive feedback mechanism.

Kaneko and Hiroshige [1978b] showed that a rise in median eminence CRF activity in rats produced by i.v. injected histamine was significantly suppressed when plasma corticosterone levels were rising at a rate of 6 μg/100 ml/min during i.v. infusion of corticosterone, suggesting that the site of the fast, rate-sensitive feedback inhibition exists at or above the level of hypothalamic CRF neurons.

Buckingham [1979] clearly demonstrated a depression of hypothalamic CRF response to stressful stimuli during a period of delayed feedback, but not during a silent period; a rise in hypothalamic CRF activity induced by exposure to ether vapor for 1 min was suppressed in rats injected i.p. with corticosterone 1 h but not 15 min previously, although plasma corticosterone levels were markedly elevated at 15 min after corticosterone injection and returned to the basal levels at 1 h.

In experiments in vitro with incubated rat hypothalami *Jones* et al. [1977] demonstrated both fast and delayed feedback effects of corticosterone added into the incubation medium on hypothalamic CRF response to acetylcholine. In the in vitro experiments of *Buckingham* [1979] with incubated hypothalami removed from rats which were injected 1 h previously with corticosterone, the release and production of hypothalamic CRF in response to acetylcholine or 5-hydroxytryptamine, but not the basal CRF activity, were found to be suppressed. Using hypothalamus-pituitary cell-adrenal cell superfusion system, *Vermes* et al. [1977] showed that the basal CRF release of the hypothalamus but not the veratridine- or electrical stimulation-induced hypothalamic CRF release was blocked in vitro by presuperfusion with dexamethasone. Suppression of basal hypothalamic CRF release in vitro was also observed in experiments with superfused hypothalami which were removed from rats injected i.p. with corticosterone 30 min previously. These experimental results suggest that the site of inhibitory feedback action of corticosteroids may be located in the hypothalamus.

C. 'Short-Loop' Negative Feedback Mechanisms Mediated by Adrenocorticotropic Hormone

Possible mechanisms of feedback inhibition by ACTH on pituitary ACTH release ('short-loop' negative feedback mechanisms) have been suggested by *Hodges and Vernikos* [1958, 1959] and *Kitay* et al. [1959a]. The former investigators showed that the rise of circulating ACTH levels in response to ether stress in adrenalectomized rats, whose resting circulatory ACTH was non-detectable 5 days after adrenalectomy and was attained to high levels on 25 or 35 days, depended on the preexisting circulating ACTH levels; it was significantly greater in rats with low than in rats with high resting levels of ACTH. The latter investigators found that pretreatment with ACTH blocked adrenal ascorbic acid depletion in intact rats in response to ether stress and pituitary ACTH depletion in adrenalectomized rats in response to scalding (immersion to the neck in water at 70 °C).

The concept of the 'short-loop' negative feedback mechanisms in the regulation of pituitary ACTH release has further been supported by hypothalamic implantation experiments. Plasma corticosterone levels in rats are significantly depressed by implantation of ACTH into the median eminence but not into the frontal cerebral cortex or into the pituitary [*Motta* et al., 1965]. Thus, it is suggested that ACTH receptors for the 'short-loop' negative feedback mechanisms might be located in the median eminence. However, as emphasized by *Seiden and Brodish* [1971] there is a possibility of 'implantation paradox' of *Bogdanove* [1963], i.e. a distribution into the entire anterior pituitary of ACTH implanted in the median eminence.

Sites and mechanisms of action of ACTH in inhibiting pituitary ACTH release have also been examined by evaluating the hypothalamic CRF activity. An increase in hypothalamic CRF activity in rats following adrenalectomy is suppressed by treatment with ACTH [*Legori* et al., 1965]. The CRF activity of the median eminence is increased by adrenalectomy and it is further increased by additional hypophysectomy; this elevated median eminence CRF activity in adrenalectomized-hypophysectomized rats is reduced by treatment with ACTH to the level in adrenalectomized rats [*Motta* et al., 1969]. An increase in median eminence CRF activity in chronically adrenalectomized rats in response to ether stress is significantly but not strongly depressed by daily pretreatment with ACTH for 4 days, but it is not reduced by ACTH injected 4 or 24 h previously [*Hiroshige* et al., 1969b]. A marked increase (3–4 times) in hypothalamic CRF activity in rats is observed 1–6 weeks but not 2–4 days after hypophysectomy and this increase in hypothalamic CRF activity in chronically hypophysectomized rats is not reduced by adrenalectomy performed 1–2 days after hypophysectomy [*Seiden and Brodish,* 1971]. In chronically hypophysectomized-adrenalectomized rats a long-term (1-week) treatment with ACTH reduces the hypothalamic CRF activity by 87% [*Seiden and Brodish,* 1971]. A significant decrease in hypothalamic CRF activity of hypophysectomized-adrenalectomized rats following a long-term treatment with ACTH has been confirmed by *Takebe* et al. [1974]. They have also observed that the hypothalamic CRF activity under ether stress but not under ether plus surgical stress (laparotomy followed by intestinal traction) is suppressed by a long-term (5-day) pretreatment with ACTH. These findings may be interpreted to mean that circulating ACTH is one of the factors which participate in the regulation of the rate of synthesis and/or release of hypothalamic CRF.

References

Abe, K.; Critchlow, V.: Effects of corticosterone, dexamethasone and surgical isolation of the medial basal hypothalamus on rapid feedback control of stress-induced corticotropin secretion in female rats. Endocrinology *101*: 498–505 (1977).

Abe, K.; Hirose, T.: Effect of corticosterone and cortisol on corticosteroidogenesis in isolated rat adrenal cells. Tohoku J. exp. Med. *112*: 195–196 (1974).

Abe, K.; Hiroshige, T.: Changes in plasma corticosterone and hypothalamic CRF levels following intraventricular injection or drug-induced changes of brain biogenic amines in the rat. Neuroendocrinology *14*: 195–211 (1974).

Abe, K.; Kroning, J.; Greer, M.A.; Critchlow, V.: Effects of destruction of the suprachiasmatic nuclei on the circadian rhythms in plasma corticosterone, body temperature, feeding and plasma thyrotropin. Neuroendocrinology *29*: 119–131 (1979).

Abelson, D.; Baron, D.N.: The effect of cortisone acetate on adrenal ascorbic-acid depletion following stress. Lancet *ii*: 663–664 (1952).

Ader, R.: Early experiences accelerate maturation of the 24-hour adrenocortical rhythm. Science *163*: 1225–1226 (1969).

Ader, R.; Friedman, S.B.: Plasma corticosterone response to environmental stimulation: effects of duration of stimulation and the 24-hour adrenocortical rhythm. Neuroendocrinology *3*: 378–386 (1968).

Ahlers, I.; Šmajda, B.; Ahlersová, E.: Circadian rhythm of plasma and adrenal corticosterone in rats: effect of restricted feeding schedules. Endocrinol. exp. *14*: 183–190 (1980).

Aikawa, T.; Hirose, T.; Matsumoto, I.; Suzuki, T.: Secretion of aldosterone in response to histamine in hypophysectomized-nephrectomized dogs. J. Endocr. *81*: 325–330 (1979).

Aikawa, T.; Hirose, T.; Matsumoto, I.; Suzuki, T.: Adrenal secretion of aldosterone in response to anaphylactic shock in hypophysectomized-nephrectomized dogs. Jap. J. Physiol. *31*: 145–151 (1981a).

Aikawa, T.; Hirose, T.; Matsumoto, I.; Suzuki, T.: Direct stimulatory effect of histamine on aldosterone secretion of the perfused dog adrenal gland. Jap. J. Physiol. *31*: 457–463 (1981b).

Allen, C.; Kendall, J.W.: Maturation of the circadian rhythm of plasma corticosterone in the rat. Endocrinology *80*: 926–930 (1967).

Allen, J.P.; Allen, C.F.: Role of the amygdaloid complexes in the stress-induced release of ACTH in the rat. Neuroendocrinology *15*: 220–230 (1974).

Allen-Rowlands, C.F.; Allen, J.P.; Greer, M.A.; Wilson, M.: Circadian rhythmicity of ACTH and corticosterone in the rat. J. Endocrinol. Invest. *4*: 371–377 (1980).

Amatruda, T.T., Jr.; Hollingsworth, D.R.; D'Esopo, N.D.; Upton, G.V.; Bondy, P.K.: A study of the mechanism of the steroid withdrawal syndrome. Evidence for integrity of the hypothalamic-pituitary-adrenal system. J. clin. Endocr. Metab. *20:* 339–354 (1960).

Andersen, R.N.; Egdahl, R.H.: Effect of vasopressin on pituitary-adrenal secretion in the dog. Endocrinology *74:* 538–542 (1964).

Andersson, K.-E.; Arner, B.; Hedner, P.; Mulder, J.L.: Effects of 8-lysine-vasopressin and synthetic analogues on release of ACTH. Acta endocr., Copenh. *69:* 640–648 (1972).

Anichkov, S.V.; Malyghina, E.I.; Poskalenko, A.N.; Ryzhenkov, V.E.: Reflexes from carotid bodies upon the adrenals. Archs int. Pharmacodyn. Thér. *129:* 156–165 (1960).

Arcangeli, P.; Digiesi, V.; Madeddu, G.; Toccafondi, R.: Temporary displacement of plasma corticoid circadian peak induced by ablation of olfactory bulbs in dog. Experientia *29:* 358–359 (1973).

Arimura, A.: The effect of posterior-pituitary hormone on the release of ACTH. Jap. J. Physiol. *5:* 37–44 (1955/56).

Arimura, A.; Bowers, C.Y.; Schally, A.V.; Saito, M.; Miller, M.C.: Effect of corticotropin-releasing factor, dexamethasone and actinomycin D on the release of ACTH from rat pituitaries in vivo and in vitro. Endocrinology *85:* 300–311 (1969).

Arimura, A.; Long, C.N.H.: Influence of a small dose of vasopressin upon the pituitary-adrenal activation in the rat. Jap. J. Physiol. *12:* 411–422 (1962).

Arimura, A.; Saito, T.; Bowers, C.Y.; Schally, A.V.: Pituitary-adrenal activation in rats with hereditary hypothalamic diabetes insipidus. Acta endocr., Copenh. *54:* 155–165 (1967).

Arimura, A.; Yamaguchi, T.; Yoshimura, K.; Imazeki, T.; Itoh, S.: Role of the neurohypophysis in the release of adrenocorticotrophic hormone in the rat. Jap. J. Physiol. *15:* 278–295 (1965).

Armbruster, H.; Vetter, W.; Beckerhoff, R.; Nussberger, J.; Vetter, H.; Siegenthaler, W.: Diurnal variations of plasma aldosterone in supine man: relationship to plasma renin activity and plasma cortisol. Acta endocr., Copenh. *80:* 95–103 (1975).

Armstrong, W.E.; Hatton, G.I.: Morphological changes in the rat supraoptic and paraventricular nuclei during the diurnal cycle. Brain Res. *157:* 407–413 (1978).

Arner, B.; Hedner, P.; Karlefors, T.: Adrenocortical activity during induced hypoglycaemia. An experimental study in man. Acta endocr., Copenh. *40:* 421–429 (1962).

Asano, H.: Effects of insulin, histamine and vasopressin on pituitary-adrenocortical secretion with or without dexamethasone pretreatment. Nagoya J. med. Sci. *29:* 139–154 (1966).

Aschan, G.: Oxygen deficiency and oxygen poisoning as stress factors. Acta Soc. Med. upsal. *58:* 265–268 (1953).

Asfeldt, V.H.; Elb, S.: Hypothalamo-pituitary-adrenal response during major surgical stress. Acta endocr., Copenh. *59:* 67–75 (1968).

Ashford, A.; Shapero, M.: Effect of chlorpromazine, reserpine, benactyzine and phenobarbitone on the release of corticotrophin in the rat. Br. J. Pharmacol. *19:* 458–463 (1962).

Atkins, G.; Marotta, S.F.: Relationship of altered vascular volumes to plasma 17-hydroxycorticosteroids in dogs. Proc. Soc. exp. Biol. Med. *113:* 461–465 (1963).

Bacq, Z.M.; Fischer, P.: The action of various drugs on the suprarenal response of the rat to total-body x-irradiation. Radiat. Res. *7:* 365–372 (1957).

Bacq, Z.M.; Smelik, P.G.; Goutier-Pirotte, M.; Renson, J.: Effect of the destruction of hypothalamus on the suprarenal response of the rat to total body irradiation. Br. J. Radiol. *33:*618–621 (1960).

Balfour, D.J.K.; Khullar, A.K.; Longden, A.: Effects of nicotine on plasma corticosterone and brain amines in stressed and unstressed rats. Pharmacol. Biochem. Behav. *3:*179–184 (1975).

Barlow, S.M.; Morrison, P.J.; Sullivan, F.M.: Effects of acute and chronic stress on plasma corticosterone levels in the pregnant and non-pregnant mouse. J. Endocr. *66:*93–99 (1975).

Barrett, A.M.; Stockham, M.A.: The response of the pituitary-adrenal system to a stressful stimulus: the effect of conditioning and pentobarbitone treatment. J. Endocr. *33:*145–152 (1965).

Beall, R.J.; Sayers, G.: Isolated adrenal cells: steroidogenesis and cyclic AMP accumulation in response to ACTH. Archs Biochem. Biophys. *148:*70–76 (1972).

Beigelman, P.M.; Slusher, M.A.; Slater, G.G.; Roberts, S.: Effect of anesthetics and collection time on corticosteroid secretion by rat adrenal. Proc. Soc. exp. Biol. Med. *93:*608–611 (1956).

Bellinger, L.L.; Bernardis, L.L.; Mendel, V.E.: Effect of ventromedial and dorsomedial hypothalamic lesions on circadian corticosterone rhythms. Neuroendocrinology *22:*216–225 (1976).

Berthold, K.; Arimura, A.; Schally, A.V.: Effect of 6-dehydro-16-methylene-hydrocortisone and dexamethasone on the release of ACTH from rat pituitary glands in vitro. Acta endocr., Copenh. *63:*431–436 (1970).

Betz, D.; Ganong, W.F.: Effect of chlorpromazine on pituitary-adrenal function in the dog. Acta endocr., Copenh. *43:*264–270 (1963).

Bhattacharya, A.N.; Marks, B.H.: Effects of pargyline and amphetamine upon acute stress responses in rats. Proc. Soc. exp. Biol. Med. *130:*1194–1198 (1969a).

Bhattacharya, A.N.; Marks, B.H.: Reserpine- and chlorpromazine-induced changes in hypothalamo-hypophyseal-adrenal system in rats in the presence and absence of hypothermia. J. Pharmac. exp. Ther. *165:*108–116 (1969b).

Bhattacharya, A.N.; Marks, B.H.: Effects of alpha methyl tyrosine and *p*-chlorophenylalanine on the regulation of ACTH secretion. Neuroendocrinology *6:*49–55 (1970).

Biddulph, C.; Finerty, J.C.; Ellis, J.P., Jr.: Blood corticosteroids and anterior pituitary ACTH and cytology of dogs exposed to hypocapnia and/or hypoxemia. Am. J. Physiol. *197:*126–128 (1959).

Binhammer, R.T.; Crocker, J.R.: Effect of X-irradiation on the pituitary-adrenal axis of the rat. Radiat. Res. *18:*429–436 (1963).

Birmingham, M.K.; Elliott, F.H.; Valère, P.H.-L.: The need for the presence of calcium for the stimulation in vitro of rat adrenal glands by adrenocorticotrophic hormone. Endocrinology *53:*687–689 (1953).

Birmingham, M.K.; Kurlents, E.: Inactivation of ACTH by isolated rat adrenals and inhibition of corticoid formation by adrenocortical hormones. Endocrinology *62:*47–60 (1958).

Black, W.C.; Crampton, R.S.; Verdesca, A.S.; Nedeljkovic, R.I.; Hilton, J.G.: Inhibitory effect of hydrocortisone and analogues on adrenocortical secretion in dogs. Am. J. Physiol. *201:*1057–1060 (1961).

Blair-West, J.R.; Coghlan, J.P.; Denton, D.A.; Funder, J.W.; Scoggins, B.A.; Wright,

R.D.: Effects of prostaglandin E₁ upon the steroid secretion of the adrenal of the sodium deficient sheep. Endocrinology 88:367–371 (1971).

Blanchard, K.C.; Dearborn, E.H.; Maren, T.H.; Marshall, E.K., Jr.: Stimulation of the anterior pituitary by certain cinchoninic acid derivatives. Bull. Johns Hopkins Hosp. 86:83–88 (1950).

Blichert-Toft, M.; Hippe, E.; Jensen, H.K.: Adrenal cortical function as reflected by the plasma hydrocortisone and urinary 17-ketogenic steroids in relation to surgery in elderly patients. Acta chir. scand. 133:591–599 (1967).

Bliss, E.L.; Migeon, C.J.; Eik-Nes, K.; Sandberg, A.A.; Samuels, L.T.: The effects of insulin, histamine, bacterial pyrogen, and the antabuse-alcohol reaction upon the levels of 17-hydroxycorticosteroids in the peripheral blood of man. Metabolism 3: 493–501 (1954).

Bliss, E.L.; Sandberg, A.A.; Nelson, D.H.; Eik-Nes, K.: The normal levels of 17-hydroxycorticosteroids in the peripheral blood of man. J. clin. Invest. 32:818–823 (1953).

Bloom, S.R.; Edwards, A.V.; Hardy, R.N.; Silver, M.: Adrenal and pancreatic endocrine responses to hypoxia in the conscious calf. J. Physiol., Lond. 261:271–283 (1976).

Boddy, K.; Jones, C.T.; Mantell, C.; Ratcliffe, J.G.; Robinson, J.S.: Changes in plasma ACTH and corticosteroid of the maternal and fetal sheep during hypoxia. Endocrinology 94:588–591 (1974).

Bogdanove, E.M.: Direct gonad-pituitary feedback: an analysis of effects of intracranial estrogenic depots on gonadotrophin secretion. Endocrinology 73:696–712 (1963).

Bohus, B.; Endröczi, E.: Untersuchungen über die Wirkung von Chlorpromazin auf das Hypophysen-Nebennierenrinden-System bei Ratten. Endokrinologie 46: 126–133 (1964a).

Bohus, B.; Endröczi, E.: Effect of intracerebral implantation of hydrocortisone on adrenocortical secretion and adrenal weight after unilateral adrenalectomy. Acta physiol. hung. 25:11–19 (1964b).

Bohus, B.; Nyakas, Cs.; Lissák, K.: Involvement of suprahypothalamic structures in the hormonal feedback action of corticosteroids. Acta physiol. hung. 34:1–8 (1968).

Bohus, B.; Strashimirov, D.: Localization and specificity of corticosteroid 'feedback receptors' at the hypothalamo-hypophyseal level; comparative effects of various steroids implanted in the median eminence or the anterior pituitary of the rat. Neuroendocrinology 6:197–209 (1970).

Borrell, J.; Llorens, I.; Borrell, S.: Study of the effects of morphine on adrenal corticosteroids, ascorbic acid and catecholamines in unanaesthetized and anaesthetized cats. Hormone Res. 5:351–358 (1974).

Božović, Lj.; Kostial-Živanović, K.: Muscular work and adrenocortical activity. Archs int. Physiol. 60:459–464 (1952).

Bradbury, M.W.B.; Burden, J.; Hillhouse, E.W.; Jones, M.T.: Stimulation electrically and by acetylcholine of the rat hypothalamus in vitro. J. Physiol., Lond. 239: 269–283 (1974).

Briggs, F.N.; Munson, P.L.: Studies on the mechanism of stimulation of ACTH secretion with the aid of morphine as a blocking agent. Endocrinology 57: 205–219 (1955).

Brodie, B.B.; Costa, E.; Dlabac, A.; Neff, N.H.; Smookler, H.H.: Application of steady state kinetics to the estimation of synthesis rate and turnover time of tissue catecholamines. J. Pharmac. exp. Ther. 154:493–498 (1966).

Brodish, A.: Diffuse hypothalamic system for the regulation of ACTH secretion. Endocrinology *73:*727–735 (1963).

Brodish, A.: Delayed secretion of ACTH in rats with hypothalamic lesions. Endocrinology *74:*28–34 (1964).

Brodish, A.: Effect of hypothalamic lesions on the time course of corticosterone secretion. Neuroendocrinology *5:*33–47 (1969).

Brostoff, J.; James, V.H.T.; Landon, J.: Plasma corticosteroid and growth hormone response to lysine-vasopressin in man. J. clin. Endocr. Metab. *28:*511–518 (1968).

Brown, G.M.; Schalch, D.S.; Reichlin, S.: Hypothalamic mediation of growth hormone
and adrenal stress response in the squirrel monkey. Endocrinology *89:* 694–703
(1971).

Brown, H.; Englert, E., Jr.; Wallach, S.; Simons, E.L.: Metabolism of free and conjugated 17-hydroxycorticosteroids in normal subjects. J. clin. Endocr. Metab. *17:*
1191–1201 (1957).

Brownie, A.C.; Kramer, R.E.; Gallant, S.: The cholesterol side chain cleavage system of
the rat adrenal cortex and its relationship to the circadian rhythm. Endocrinology
*104:*1266–1269 (1979).

Buckingham, J.C.: The influence of corticosteroids on the secretion of corticotrophin
and its hypothalamic releasing hormone. J. Physiol., Lond. *286:*331–342 (1979).

Buckingham, J.C.; Hodges, J.R.: Hypothalamo-pituitary adrenocortical function in the
rat after treatment with betamethasone. Br. J. Pharmacol. *56:*235–239 (1976).

Buckingham, J.C.; Hodges, J.R.: The use of corticotrophin production by adenohypophysial tissue in vitro for the detection and estimation of potential corticotrophin
releasing factors. J. Endocr. *72:*187–193 (1977).

Butte, J.C.; Kakihana, R.; Noble, E.P.: Circadian rhythm of corticosterone levels in rat
brain. J. Endocr. *68:*235–239 (1976).

Cade, R.; Shires, D.L.; Barrow, M.V.; Thomas, W.C., Jr.: Abnormal diurnal variation of
plasma cortisol in patients with renovascular hypertension. J. clin. Endocr. Metab.
*27:*800–806 (1967).

Campbell, C.B.G.; Ramaley, J.A.: Retinohypothalamic projections: correlations
with onset of the adrenal rhythm in infant rats. Endocrinology *94:* 1201–1204
(1974).

Cann, M.C.; Zawidzka, Z.Z.; Airth, J.M.; Grice, H.C.: The effect of ether anesthesia on
plasma corticosteroids and hematologic responses. Can. J. Physiol. Pharmacol. *43:*
463–468 (1965).

Carr, L.A.; Moore, K.E.: Effects of reserpine and α-methyltyrosine on brain catecholamines and the pituitary-adrenal response to stress. Neuroendocrinology *3:*
285–302 (1968).

Carsia, R.V.; Malamed, S.: Acute self-suppression of corticosteroidogenesis in isolated
adrenocortical cells. Endocrinology *105:*911–914 (1979).

Casady, R.L.; Taylor, A.N.: Effect of electrical stimulation of the hippocampus upon
corticosteroid levels in the freely-behaving, non-stressed rat. Neuroendocrinology
*20:*68–78 (1976).

Casentini, S.; De Poli, A.; Hukovic, S.; Martini, L.: Studies on the control of corticotrophin release. Endocrinology *64:*483–493 (1959).

Casentini, S.; De Poli, A.; Martini, L.: Hypothalamo-neurohypophysial involvement in
the corticotrophic action of acetylcholine. Br. J. Pharmacol. *12:*166–170 (1957).

Challis, J.R.G.; Carson, G.D.; Naftolin, F.: Effect of prostaglandin E_2 on the concentration of cortisol in the plasma of newborn lambs. J. Endocr. 76: 177–178 (1978).

Chambers, J.W.; Brown, G.M.: Neurotransmitter regulation of growth hormone and ACTH in the rhesus monkey: effects of biogenic amines. Endocrinology 98: 420–428 (1976).

Chan, L.T.; De Wied, D.; Saffran, M.: Comparison of assays for corticotrophin-releasing activity. Endocrinology 84: 967–972 (1969).

Chauvet, J.; Acher, R.: Influence de la vasopressine sur la sécrétion de la corticotropine (ACTH). Annls Endocr. 20: 111–115 (1959).

Cheifetz, P.; Gaffud, N.; Dingman, J.F.: Effects of bilateral adrenalectomy and continuous light on the circadian rhythm of corticotropin in female rats. Endocrinology 82: 1117–1124 (1968).

Cheifetz, P.N.; Gaffud, N.T.; Dingman, J.F.: The effect of lysine vasopressin and hypothalamic extracts on the rate of corticosterone secretion in rats treated with dexamethasone and pentobarbitone. J. Endocr. 43: 521–528 (1969).

Cheymol, J.; De Leeuw, J.; Oger, J.: Que faut-il penser de l'hypophysectomie pharmacodynamique par la chlorpromazine? C. r. Séanc. Soc. Biol. 148: 1213–1216 (1954).

Chin, A.K.; Evonuk, E.: Changes in plasma catecholamine and corticosterone levels after muscular exercise. J. appl. Physiol. 30: 205–207 (1971).

Chowers, I.; Conforti, N.; Feldman, S.: Effects of corticosteroids on hypothalamic corticotropin releasing factor and pituitary ACTH content. Neuroendocrinology 2: 193–199 (1967).

Chowers, I.; Conforti, N.; Feldman, S.: Body temperature and adrenal function in cold-exposed hypothalamic-disconnected rats. Am. J. Physiol. 223: 341–345 (1972).

Chowers, I.; Conforti, N.; Siegel, R.A.; Feldman, S.: Body temperature and adrenal function in heat-exposed hypothalamic disconnected rats. Pflügers Arch. 363: 245–250 (1976).

Chowers, I.; Einat, R.; Feldman, S.: Effects of starvation on levels of corticotrophin releasing factor, corticotrophin and plasma corticosterone in rats. Acta endocr., Copenh. 61: 687–694 (1969).

Chowers, I.; Feldman, S.; Davidson, J.M.: Effects of intrahypothalamic crystalline steroids on acute ACTH secretion. Am. J. Physiol. 205: 671–673 (1963).

Chowers, I.; Hammel, H.T.; Eisenman, J.; Abrams, R.M.; McCann, S.M.: Comparison of effect of environmental and preoptic heating and pyrogen on plasma cortisol. Am. J. Physiol. 210: 606–610 (1966).

Chowers, I.; Hammel, H.T.; Stromme, S.B.; McCann, S.M.: Comparison of effect of environmental and preoptic cooling on plasma cortisol levels. Am. J. Physiol. 207: 577–582 (1964).

Chowers, I.; Siegel, R.; Conforti, N.; Feldman, S.: Effect of acute neurogenic stress on hypothalamic corticotropin-releasing factor content. Israel J. Med. Scis. 9: 1056–1058 (1973).

Christy, N.P.; Donn, A.; Jailer, J.W.: Inhibition by aminopyrine of adrenocortical activation caused by pyrogenic reaction. Proc. Soc. exp. Biol. Med. 91: 453–456 (1956).

Christy, N.P.; Longson, D.; Horwitz, W.A.; Knight, M.M.: Inhibitory effect of chlorpromazine upon the adrenal cortical response to insulin hypoglycemia in man. J. clin. Invest. 36: 543–549 (1957).

Clayton, G.W.; Librik, L.; Gardner, R.L.; Guillemin, R.: Studies on the circadian rhythm of pituitary adrenocorticotropic release in man. J. clin. Endocr. Metab. *23:* 975-980 (1963).

Colucci, C.F.; D'Alessandro, B.; Bellastella, A.; Montalbetti, N.: Circadian rhythm of plasma cortisol in the aged (Cosinor method). Gerontol. clin. *17:* 89-95 (1975).

Conlee, R.K.; Rennie, M.J.; Winder, W.W.: Skeletal muscle glycogen content: diurnal variation and effects of fasting. Am. J. Physiol. *231:* 614-618 (1976).

Conte-Devolx, B.; Oliver, C.; Giraud, P.; Gillioz, P.; Castanas, E.; Lissitzky, J.-C.; Boudouresque, F.; Millet, Y.: Effect of nicotine on in vivo secretion of melanocorticotropic hormones in the rat. Life Sci. *28:* 1067-1073 (1981).

Cook, D.M.; Kendall, J.W.; Greer, M.A.; Kramer, R.M.: The effect of acute or chronic ether stress on plasma ACTH concentration in the rat. Endocrinology *93:* 1019-1024 (1973).

Coover, G.D.; Goldman, L.; Levine, S.: Plasma corticosterone levels during extinction of a lever-press response in hippocampectomized rats. Physiol. Behav. *7:* 727-732 (1971).

Corbin, A.; Mangili, G.; Motta, M.; Martini, L.: Effect of hypothalamic and mesencephalic steroid implantations on ACTH feedback mechanisms. Endocrinology *76:* 811-818 (1965).

Corcoran, A.C.; Page, I.H.: Methods for the chemical determination of corticosteroids in urine and plasma. J. Lab. clin. Med. *33:* 1326-1333 (1948).

Cornil, A.; De Coster, A.; Copinschi, G.; Franckson, J.R.M.: Effect of muscular exercise on the plasma level of cortisol in man. Acta endocr., Copenh. *48:* 163-168 (1965).

Coste, F.; Bourel, M.; Delbarre, F.: Déplétion ascorbique surrénale sous l'influence du salicylate de sodium chez le rat hypophysectomisé. C. r. Séanc. Soc. Biol. *147:* 668-670 (1953).

Crabbé, J.; Riondel, A.; Mach, E.: Contribution à l'étude des réactions corticosurrénaliennes. Modifications du taux des 17-OH corticostéroïdes plasmatiques à la suite d'un effort physique (compétition d'aviron). Acta endocr., Copenh. *22:* 119-124 (1956).

Craig, B.W.; Griffith, D.R.: Effects of thyroxine and exercise on the glandular and plasma levels of corticosterone in the male rat. Experientia *32:* 939-940 (1976).

Critchlow, V.: The role of light in the neuroendocrine system; in Nalbandov, Advances in neuroendocrinology, pp. 377-426 (University of Illinois Press, Urbana 1963).

Critchlow, V.; Liebelt, R.A.; Bar-Sela, M.; Mountcastle, W.; Lipscomb, H.S.: Sex difference in resting pituitary-adrenal function in the rat. Am. J. Physiol. *205:* 807-815 (1963).

Cronheim, G.; King, J.S., Jr.; Hyder, N.: Effect of salicylic acid and similar compounds on the adrenal-pituitary system. Proc. Soc. exp. Biol. Med. *80:* 51-55 (1952).

Cuello, A.C.; Scapagnini, U.; Ličko, V.; Preziosi, P.; Ganong, W.F.: Effect of dihydroxyphenylserine on the increase in plasma corticosterone in rats treated with α-methyl-p-tyrosine. Neuroendocrinology *13:* 115-122 (1973/74).

Cuello, A.C.; Shoemaker, W.J.; Ganong, W.F.: Effect of 6-hydroxydopamine on hypothalamic norepinephrine and dopamine content, ultrastructure of the median eminence, and plasma corticosterone. Brain Res. *78:* 57-69 (1974).

Cugini, P.; Giovannini, C.; Rossi, G.; Sellini, M.; Sartori, M.P.: Influenza del propranololo sulle variazioni diarie della cortisolemia nell'uomo. Boll. Soc. ital. Biol. sper. *51:* 442-447 (1975).

Cugini, P.; Manconi, R.; Serdoz, R.; Mancini, A.; Meucci, T.: Influence of propranolol on circadian rhythms of plasma renin, aldosterone and cortisol in healthy supine man. Boll. Soc. ital. Biol. sper. *53:*263–269 (1977).

Czaja, C.; Kalant, H.: The effect of acute alcoholic intoxication on adrenal ascorbic acid and cholesterol in the rat. Can. J. Biochem. Physiol. *39:*327–334 (1961).

Daiguji, M.; Mikuni, M.; Okada, F.; Yamashita, I.: The diurnal variations of dopamine-β-hydroxylase activity in the hypothalamus and locus coeruleus of the rat. Brain Res. *155:*409–412 (1978).

Daily, W.J.R.; Ganong, W.F.: The effect of ventral hypothalamic lesions on sodium and potassium metabolism in the dog. Endocrinology *62:*442–454 (1958).

Dallman, M.F.; Engeland, W.C.; Rose, J.C.; Wilkinson, C.W.; Shinsako, J.; Siedenburg, F.: Nycthemeral rhythm in adrenal responsiveness to ACTH. Am. J. Physiol. *235:* R210–R218 (1978).

Dallman, M.F.; Jones, M.T.: Corticosteroid feedback control of ACTH secretion: effect of stress-induced corticosterone secretion on subsequent stress responses in the rat. Endocrinology *92:* 1367–1375 (1973).

Dallman, M.F.; Yates, F.E.: Characteristics of corticosteroid inhibition of CRF secretion. Fed. Proc. *26:*315 (1967).

Dallman, M.F.; Yates, F.E.: Dynamic asymmetries in the corticosteroid feedback path and distribution-metabolism-binding elements of the adrenocortical system. Ann. N.Y. Acad. Sci. *156:*696–721 (1969).

David-Nelson, M.A.; Brodish, A.: Evidence for a diurnal rhythm of corticotrophin-releasing factor (CRF) in the hypothalamus. Endocrinology *85:* 861–866 (1969).

Davidson, J.M.; Feldman, S.: Cerebral involvement in the inhibition of ACTH secretion by hydrocortisone. Endocrinology *72:*936–946 (1963).

Davidson, J.M.; Feldman, S.: Effects of extrahypothalamic dexamethasone implants on the pituitary-adrenal system. Acta endocr., Copenh. *55:*240–246 (1967).

Davidson, J.M.; Jones, L.E.; Levine, S.: Feedback regulation of adrenocorticotropin secretion in 'basal' and 'stress' conditions: acute and chronic effects of intrahypothalamic corticoid implantation. Endocrinology *82:*655–663 (1968).

Davies, C.T.M.; Few, J.D.: Effect of hypoxia on the adrenocortical response to exercise in man. J. Endocr. *71:*157–158 (1976).

Davis, J.O.; Anderson, E.; Carpenter, C.C.J.; Ayers, C.R.; Haymaker, W.; Spence, W.T.: Aldosterone and corticosterone secretion following midbrain transection. Am. J. Physiol. *200:*437–443 (1961).

Degli Uberti, E.; Trasforini, G.; Margutti, A.; Rotola, C.; Lo Vecchio, G.: Failure of metergoline to affect the circadian periodicity of plasma cortisol levels in healthy man. Boll. Soc. ital. Biol. sper. *57:*367–373 (1981).

Degonskii, A.I.: The hypothalamo-adenohypophyseo-adrenal neurosecretory system in hyperthermia. Bull. Exp. Biol. Med. *75:*253–255 (1973).

Degonskii, I.A.: Role of adrenergic structures of the CNS in changes in function of the hypothalamic-pituitary-adrenal system during exogenous hyperthermia. Bull. Exp. Biol. Med. *83:*293–295 (1977).

De Souza, E.B.; Van Loon, G.R.: Stress-induced inhibition of the plasma corticosterone response to a subsequent stress in rats: a nonadrenocorticotropin-mediated mechanism. Endocrinology *110:*23–33 (1982).

De Wied, D.: The site of the blocking action of dexamethasone on stress-induced pituitary ACTH release. J. Endocr. *29*: 29–37 (1964).

De Wied, D.: Influence of vasopressin and of a crude CRF preparation on pituitary ACTH-release in posterior-lobectomized rats. Neuroendocrinology *3:* 129–135 (1968).

De Wied, D.; Bouman, P.R.; Smelik, P.G.: The effect of a lipide extract from the posterior hypothalamus and of Pitressin on the release of ACTH from the pituitary gland. Endocrinology *62:* 605–613 (1958).

De Wied, D.; Mirsky, I.A.: The action of Δ¹ hydrocortisone on the antidiuretic and adrenocorticotropic responses to noxious stimuli. Endocrinology *64:* 955–966 (1959).

De Wied, D.; Siderius, P.; Mirsky, I.A.: The antidiuretic- and ACTH-releasing effect of various octapeptides. Archs int. Pharmacodyn. Thér. *133:* 50–57 (1961).

De Wied, D.; Witter, A.; Versteeg, D.H.G.; Mulder, A.H.: Release of ACTH by substances of central nervous system origin. Endocrinology *85:* 561–569 (1969).

Dixit, B.N.; Buckley, J.P.: Brain 5-hydroxytryptamine and anterior pituitary activation by stress. Neuroendocrinology *4:* 32–41 (1969).

Doe, R.P.; Flink, E.B.; Goodsell, M.G.: Relationship of diurnal variation in 17-hydroxycorticosteroid levels in blood and urine to eosinophils and electrolyte excretion. J. clin. Endocr. Metab. *16:* 196–206 (1956).

Dordoni, F.; Fortier, C.: Effect of eserine and atropine on ACTH release. Proc. Soc. exp. Biol. Med. *75:* 815–816 (1950).

Douglas, W.W.; Poisner, A.M.: On the mode of action of acetylcholine in evoking adrenal medullary secretion: increased uptake of calcium during the secretory response. J. Physiol., Lond. *162:* 385–392 (1962).

Douglas, W.W.; Rubin, R.P.: The role of calcium in the secretory response of the adrenal medulla to acetylcholine. J. Physiol., Lond. *159:* 40–57 (1961).

Dunn, J.; Bennett, M.; Peppler, R.: Pituitary-adrenal function in photic and olfactory deprived rats. Proc. Soc. exp. Biol. Med. *140:* 755–758 (1972).

Dunn, J.; Critchlow, V.: Feedback suppression of 'non-stress' pituitary-adrenal function in rats with forebrain removed. Neuroendocrinology *4:* 296–308 (1969).

Dunn, J.; Critchlow, V.: Vasopressin-evoked ACTH release in rats following forebrain removal. Proc. Soc. exp. Biol. Med. *136:* 1284–1288 (1971).

Dunn, J.; Critchlow, V.: Electrically stimulated ACTH release in pharmacologically blocked rats. Endocrinology *93:* 835–842 (1973).

Dunn, J.D.; Carrillo, A.J.: Circadian variation in the sensitivity of the pituitary-adrenal system to dexamethasone suppression. J. Endocr. *76:* 63–66 (1978).

Dunn, J.D.; Hess, M.; Johnson, D.C.: Effect of thyroidectomy on rhythmic gonadotropin release. Proc. Soc. exp. Biol. Med. *151:* 22–27 (1976a).

Dunn, J.D.; Johnson, D.C.: Effect of adrenalectomy on temporal patterns of serum gonadotropin levels. Neuroendocrinology *27:* 126–135 (1978).

Dunn, J.D.; Kastin, A.J.; Carrillo, A.J.: Circadian variation in plasma melanocyte stimulating hormone activity in the rat. Int. J. Chronobiol. *4:* 163–169 (1976b).

Dury, A.: Changes in circulating eosinophiles and adrenal ascorbic acid concentration after agents altering blood sugar levels and after surgical conditions. Am. J. Physiol. *163:* 96–103 (1950).

Dzieniszewski, J.; Milewski, B.: Circadian rhythm of cortisolaemia in patients with chronic aggressive hepatitis. Pol. Arch. Med. Wewn. *52:* 239–247 (1974).

Earp, H.S.; Watson, B.S.; Ney, R.L.: Adenosine 3′,5′-monophosphate as the mediator of ACTH-induced ascorbic acid depletion in the rat adrenal. Endocrinology 87: 118–123 (1970).

Edwardson, J.A.; Bennett, G.W.: Modulation of corticotrophin-releasing factor release from hypothalamic synaptosomes. Nature, Lond. 251:425–427 (1974).

Eechaute, W.; Demeester, G.; Leusen, I.: Irradiation totale aux rayons X et activité surrénalienne. Archs int. Pharmacodyn. Thér. 135:235–248 (1962a).

Eechaute, W.; Lacroix, E.; Leusen, I.; Bouckaert, J.J.: L'activité du cortex surrénalien sous l'influence de la réserpine et de l'iproniazide. Archs int. Pharmacodyn. Thér. 139:403–414 (1962b).

Egdahl, R.H.: The differential response of the adrenal cortex and medulla to bacterial endotoxin. J. clin. Invest. 38:1120–1125 (1959).

Egdahl, R.H.: Adrenal cortical and medullary responses to trauma in dogs with isolated pituitaries. Endocrinology 66:200–216 (1960a).

Egdahl, R.H.: The effect of brain removal, decortication and mid brain transection on adrenal cortical function in dogs. Acta endocr., Copenh. 35: suppl. 51, pp. 49–50 (1960b).

Egdahl, R.H.: Corticosteroid secretion following caval constriction in dogs with isolated pituitaries. Endocrinology 68:226–231 (1961a).

Egdahl, R.H.: Cerebral cortical inhibition of pituitary-adrenal secretion. Endocrinology 68:574–581 (1961b).

Egdahl, R.H.: Further studies on adrenal cortical function in dogs with isolated pituitaries. Endocrinology 71:926–935 (1962).

Egdahl, R.H.: The acute effects of steroid administration on pituitary adrenal secretion in the dog. J. clin. Invest. 43:2178–2184 (1964).

Egdahl, R.H.: Studies on the effect of ether and pentobarbital anesthesia on pituitary adrenal function in the dog. Neuroendocrinology 1:184–191 (1965/66).

Egdahl, R.H.; Melby, J.C.; Spink, W.W.: Adrenal cortical and body temperature responses to repeated endotoxin administration. Proc. Soc. exp. Biol. Med. 101: 369–372 (1959).

Egdahl, R.H.; Nelson, D.H.; Hume, D.M.: Effect of hypothermia on 17-hydroxycorticosteroid secretion in adrenal venous blood in the dog. Science 121: 506–507 (1955).

Egdahl, R.H.; Richards, J.B.: Effect of chlorpromazine on pituitary ACTH secretion in the dog. Am. J. Physiol. 185:235–238 (1956a).

Egdahl, R.H.; Richards, J.B.: Effect of extreme cold exposure on adrenocortical function in the unanesthetized dog. Am. J. Physiol. 185:239–242 (1956b).

Egdahl, R.H.; Richards, J.B.; Hume, D.M.: Effect of reserpine on adrenocortical function in unanesthetized dogs. Science 123:418 (1956).

Ehle, A.L.; Mason, J.W.; Pennington, L.L.: Plasma growth hormone and cortisol changes following limbic stimulation in conscious monkeys. Neuroendocrinology 23:52–60 (1977).

Eik-Nes, K.; Clark, L.D.: Diurnal variation of plasma 17-hydroxycorticosteroids in subjects suffering from severe brain damage. J. clin. Endocr. Metab. 18: 764–768 (1958).

Eleftheriou, B.E.; Zolovick, A.J.; Pearse, R.: Effect of amygdaloid lesions on pituitary-adrenal axis in the deermouse. Proc. Soc. exp. Biol. Med. 122:1259–1262 (1966).

Ellis, E.F.; Shen, J.C.; Schrey, M.P.; Carchman, R.A.; Rubin, R.P.: Prostacyclin: a potent stimulator of adrenal steroidogenesis. Prostaglandins *16:*483–490 (1978).

Ellis, F.W.: Effect of ethanol on plasma corticosterone levels. J. Pharmac. exp. Ther. *153:* 121–127 (1966).

Endröczi, E.; Bata, G.; Martin, J.: Untersuchungen über die Sekretion von Nebennieren-rindenhormonen. Endokrinologie *35:*280–290 (1958).

Endröczi, E.; Kovács, S.; Lissák, K.: Die Wirkung der Hypothalamusreizung auf das endokrine und somatische Verhalten. Endokrinologie *33:*271–278 (1956).

Endröczi, E.; Lissâk, K.: Interrelations between palaeocortical activity and pituitary-adrenocortical function. Acta physiol. hung. *21:*257–263 (1962).

Endröczi, E.; Lissâk, K.: Effect of hypothalamic and brain stem structure stimulation on pituitary-adrenocortical function. Acta physiol. hung. *24:*67–77 (1963).

Endröczi, E.; Lissák, K.; Bohus, B.; Kovács, S.: The inhibitory influence of archicortical structures on pituitary-adrenal function. Acta physiol. hung. *16:*17–22 (1959).

Endröczi, E.; Lissák, K.; Tekeres, M.: Hormonal 'feed-back' regulation of pituitary-adrenocortical activity. Acta physiol. hung. *18:*291–299 (1961).

Endröczi, E.; Mess, B.: Einfluss von Hypothalamusläsionen auf die Funktion des Hypo-physen-Nebennierenrinden-Systems. Endokrinologie *33:*1–8 (1955).

Endröczi, E.; Nyakas, C.: Effect of septal lesion on exploratory activity, passive avoidance learning and pituitary-adrenal function in the rat. Acta physiol. hung. *39:* 351–360 (1971).

Endröczi, E.; Schreiberg, G.; Lissák, K.: The role of central nervous activating and inhibitory structures in the control of pituitary-adrenocortical function. Effects of intracerebral cholinergic and adrenergic stimulation. Acta physiol. hung. *24:* 211–221 (1963).

Engeland, W.C.; Shinsako, J.; Winget, C.M.; Vernikos-Danellis, J.; Dallman, M.F.: Circadian patterns of stress-induced ACTH secretion are modified by corticosterone responses. Endocrinology *100:*138–147 (1977).

Espiner, E.A.; Livesey, J.H.; Ross, J.; Donald, R.A.: Dynamics of cyclic adenosine 3',5'-monophosphate release during adrenocortical stimulation in vivo. Endocrinology *95:*838–846 (1974).

Estep, H.L.; Island, D.P.; Ney, R.L.; Liddle, G.W.: Pituitary-adrenal dynamics during surgical stress. J. clin. Endocr. Metab. *23:*419–425 (1963).

Farese, R.V.; Linarelli, L.G.; Glinsmann, W.H.; Ditzion, B.R.; Paul, M.I.; Pauk, G.L.: Persistence of the steroidogenic effect of adenosine-3',5'-monophosphate in vitro: evidence for a third factor during the steroidogenic effect of ACTH. Endocrinology *85:*867–874 (1969).

Fatranská, M.; Vargová, M.; Rosival, L.; Bátora, V.; Németh, Š.; Janeková, D.: Circadian susceptibility rhythms to some organophosphate compounds in the rat. Chronobiologia *5:*39–44 (1978).

Fehm, H.L.; Voigt, K.H.; Lang, R.: Characterization of the CRF-like-activity of vasopressin in vitro. Acta endocr., Copenh. suppl. *193:* 129 (1975).

Fekete, Gy.; Görög, P.: The inhibitory action of natural and synthetic glucocorticoids on adrenal steroidogenesis at the adrenal level. J. Endocr. *27:*123–126 (1963).

Feldman, S.; Conforti, N.: Inhibition and facilitation of feedback influences of dexamethasone on adrenocortical responses to ether stress in rats with hypothalamic deafferentations and brain lesions. Acta endocr., Copenh. *82:*785–791 (1976a).

Feldman, S.; Conforti, N.: Adrenocortical responses to olfactory stimulation in rats with hypothalamic islands. J. Endocr. *69:*165–166 (1976b).

Feldman, S.; Conforti, N.: Adrenocortical responses to olfactory stimulation in rats with partial hypothalamic deafferentations and lesions. Neuroendocrinology *24:* 162–168 (1977).

Feldman, S.; Conforti, N.; Chowers, I.: Effects of partial hypothalamic deafferentations on adrenocortical responses. Acta endocr., Copenh. *69:*526–530 (1972a).

Feldman, S.; Conforti, N.; Chowers, I.: Neural pathways mediating adrenocortical responses to photic and acoustic stimuli. Neuroendocrinology *10:* 316–323 (1972b).

Feldman, S.; Conforti, N.; Chowers, I.: Effect of dexamethasone on adrenocortical responses in intact and hypothalamic deafferented rats. Acta endocr., Copenh. *73:* 660–664 (1973).

Feldman, S.; Conforti, N.; Chowers, I.: Complete inhibition of adrenocortical responses following sciatic nerve stimulation in rats with hypothalamic islands. Acta endocr., Copenh. *78:*539–544 (1975a).

Feldman, S.; Conforti, N.; Chowers, I.: Adrenocortical responses following sciatic nerve stimulation in rats with partial hypothalamic deafferentations. Acta endocr., Copenh. *80:*625–629 (1975b).

Feldman, S.; Conforti, N.; Chowers, I.; Davidson, J.M.: Differential effects of hypothalamic deafferentation on responses to different stresses. Israel J. med. Scis *4:* 908–910 (1968).

Feldman, S.; Conforti, N.; Chowers, I.; Davidson, J.M.: Pituitary-adrenal activation in rats with medial basal hypothalamic islands. Acta endocr., Copenh. *63:* 405–414 (1970).

Feldman, S.; Conforti, N.; Davidson, J.M.: Adrenocortical responses in rats with corticosteroid and reserpine implants. Neuroendocrinology *1:*228–239 (1965/66).

Feldman, S.; Conforti, N.; Davidson, J.M.: Long-term effects of intracerebral corticoid implants. Acta endocr., Copenh. *55:*440–450 (1967).

Ferin, M.; Antunes, J.L.; Zimmerman, E.; Dyrenfurth, I.; Frantz, A.G.; Robinson, A.; Carmel, P.W.: Endocrine function in female rhesus monkeys after hypothalamic disconnection. Endocrinology *101:*1611–1620 (1977).

Few, J.D.: Effect of exercise on the secretion and metabolism of cortisol in man. J. Endocr. *62:*341–353 (1974).

Few, J.D.; Gawel, M.J.; Imms, F.J.; Tiptaft, E.M.: The influence of the infusion of noradrenaline on plasma cortisol levels in man. J. Physiol., Lond. *309:*375–389 (1980).

Fichman, M.P.; Littenburg, G.; Brooker, G.; Horton, R.: Effect of prostaglandin A_1 on renal and adrenal function in man. Circulation Res. *31:*suppl. 2, pp. 19–35 (1972).

Fiore-Donati, L.; Pollice, L.; Chieco-Bianchi, L.: Response of adrenal and preputial glands of rats to administration of 5-hydroxytryptamine. Experientia *15:* 193–195 (1959).

Fischer, P.; Renson, J.; Ciccarone, P.: Effets de la 5-hydroxytryptamine sur la surrénale du rat normal, morphiné ou nouveau-né. Archs int. Physiol. Biochim. *67:* 147 (1959).

Fisher, J.D.; De Salva, S.J.: Plasma corticosterone and adrenal ascorbic acid levels in adeno- and neurohypophysectomized rats given epinephrine postoperatively. Am. J. Physiol. *197:*1263–1264 (1959).

Flack, J.D.; Jessup, R.; Ramwell, P.W.: Prostaglandin stimulation of rat corticosteroido-genesis. Science *163*:691–692 (1969).

Flack, J.D.; Ramwell, P.W.: A comparison of the effects of ACTH, cyclic AMP, dibutyryl cyclic AMP, and PGE₂ on corticosteroidogenesis in vitro. Endocrinology *90*: 371–377 (1972).

Fleischer, N.; Donald, R.A.; Butcher, R.W.: Involvement of adenosine 3′,5′-monophosphate in release of ACTH. Am. J. Physiol. *217*:1287–1291 (1969).

Fleischer, N.; Rawls, W.E.: ACTH synthesis and release in pituitary monolayer culture: effect of dexamethasone. Am. J. Physiol. *219*:445–448 (1970).

Fleischer, N.; Vale, W.: Inhibition of vasopressin-induced ACTH release from the pituitary by glucocorticoids in vitro. Endocrinology *83*:1232–1236 (1968).

Follenius, M.; Brandenberger, G.: Influence de l'exercice musculaire sur l'évolution de la cortisolémie et de la glycémie chez l'homme. Eur. J. appl. Physiol. *33*:23–33 (1974).

Forbes, J.C.; Duncan, G.M.: The effect of acute alcohol intoxication on the adrenal glands of rats and guinea pigs. Q. Jl Stud. Alcohol *12*:355–359 (1951).

Forbes, J.C.; Duncan, G.M.: Effect of intraperitoneal administration of alcohol on the adrenal levels of cholesterol and ascorbic acid in rats and guinea pigs. Q. Jl Stud. Alcohol *14*:19–21 (1953).

Fortier, C.: Nervous control of ACTH secretion; in Harris, Donovan, The pituitary gland, vol. 2, pp. 195–234 (Butterworth, London 1966).

Fortier, C.; Harris, G.W.; McDonald, I.R.: The effect of pituitary stalk section on the adrenocortical response to stress in the rabbit. J. Physiol., Lond. *136*:344–363 (1957).

Foss, M.L.; Barnard, R.J.; Tipton, C.M.: Free 11-hydroxycorticosteroid levels in working dogs as affected by exercise training. Endocrinology *89*:96–104 (1971).

Frank, H.A.; Frank, E.D.; Korman, H.; Macchi, I.A.; Hechter, O.: Corticosteroid output and adrenal blood flow during hemorrhagic shock in the dog. Am. J. Physiol. *182*: 24–28 (1955).

Frankel, R.J.; Jenkins, J.S.; Wright, J.J.: Pituitary-adrenal response to stimulation of the limbic system and lateral hypothalamus in the rhesus monkey (*Macacca mulatta*). Acta endocr., Copenh. *88*:209–216 (1978).

Franks, R.C.: Diurnal variation of plasma 17-hydroxycorticosteroids in children. J. clin. Endocr. Metab. *27*:75–78 (1967).

Franksson, C.; Gemzell, C.A.: Blood levels of 17-hydroxycorticosteroids in surgery and allied conditions. Acta chir. scand. *106*:24–30 (1953).

Fraschini, F.; Mangili, G.; Motta, M.; Martini, L.: Midbrain and feedback control of adrenocorticotrophin secretion. Endocrinology *75*:765–769 (1964).

French, A.B.; Migeon, C.J.; Samuels, L.T.; Bowers, J.Z.: Effects of whole body x-irradiation on 17-hydroxycorticosteroid levels, leucocytes and volume of packed red cells in the rhesus monkey. Am. J. Physiol. *182*:469–476 (1955).

Frenkl, R.; Csalay, L.; Csákváry, G.: Further experimental results concerning the relationship of muscular exercise and adrenal function. Endokrinologie *66*: 285–291 (1975).

Froesch, R.: Die Funktion der Nebennierenrinde in der Insulingegenregulation. Schweiz. med. Wschr. *85*:121–127 (1955).

Froesch, R.; Kägi, H.R.; Labhart, A.: Der Einfluss von Hypoglykämie und von Muskelarbeit auf die Blutkonzentration der Nebennierenrindenhormone. Schweiz. med. Wschr. *84*:304–305 (1954).

Fujieda, K.; Hiroshige, T.: Changes in rat hypothalamic content of corticotrophin-releasing factor (CRF) activity, plasma ACTH and corticosterone under stress and the effect of cycloheximide. Acta endocr., Copenh. 89: 10–19 (1978).

Fujii, K.: Changes in plasma free fatty acids after the administration of several hormones related to the pituitary-adrenocortical and the sympathetico-medullary system. Folia endocr. jap. 48: 293–294 (1972).

Fukuda, H.; Greer, M.A.: The effect of basal hypothalamic deafferentation on the nyctohemeral rhythm of plasma TSH. Endocrinology 97: 749–752 (1975).

Fukuda, H.; Greer, M.A.; Roberts, L.; Allen, C.F.; Critchlow, V.; Wilson, M.: Nyctohemeral and sex-related variations in plasma thyrotropin, thyroxine, and triiodothyronine. Endocrinology 97: 1424–1431 (1975).

Fukui, S.; Takeuchi, K.; Watanabe, F.; Kumagai, A.; Yano, S.; Nishino, K.: Influences of some steroids on the corticosterone production by rat adrenal in vitro. Endocr. jap. 8: 43–49 (1961).

Fulkerson, W.J.; Tang, B.Y.: Ultradian and circadian rhythms in the plasma concentration of cortisol in sheep. J. Endocr. 81: 135–141 (1979).

Fuller, R.W.; Snoddy, H.D.; Molloy, B.B.: Potentiation of the L-5-hydroxytryptophan-induced elevation of plasma corticosterone levels in rats by a specific inhibitor of serotonin uptake. Res. Commun. chem. Pathol. Pharmacol. 10: 193–196 (1975).

Fuller, R.W.; Snoddy, H.D.; Molloy, B.B.: Pharmacologic evidence for a serotonin neural pathway involved in hypothalamus-pituitary-adrenal function in rats. Life Sci. 19: 337–346 (1976).

Galicich, J.H.; Halberg, F.; French, L.A.: Circadian adrenal cycle in C mice kept without food and water for a day and a half. Nature, Lond. 197: 811–813 (1963).

Gallagher, T.F.; Yoshida, K.; Roffwarg, H.D.; Fukushima, D.K.; Weitzman, E.D.; Hellman, L.: ACTH and cortisol secretory patterns in man. J. clin. Endocr. Metab. 36: 1058–1068 (1973).

Gallant, S.; Brownie, A.C.: The in vivo effect of indomethacin and prostaglandin E_2 on ACTH and DBCAMP-induced steroidogenesis in hypophysectomized rats. Biochem. biophys. Res. Commun. 55: 831–836 (1973).

Gallant, S.; Brownie, A.C.: Serum corticosteroids at the high and low points of the circadian rhythm in rats with regenerating adrenals. Life Sci. 24: 1097–1102 (1979).

Gann, D.S.: Multiple pathways regulating the adrenal cortical response to hemorrhage. Fed. Proc. 25: 552 (1966).

Gann, D.S.; Cryer, G.L.; Pirkle, J.C., Jr.: Physiological inhibition and facilitation of adrenocortical response to hemorrhage. Am. J. Physiol. 232: R5–R9 (1977).

Gann, D.S.; Egdahl, R.H.: Responses of adrenal corticosteroid secretion to hypotension and hypovolemia. J. clin. Invest. 44: 1–7 (1965).

Ganong, W.F.: Neurotransmitters involved in ACTH secretion: catecholamines. Ann. N.Y. Acad. Sci. 297: 509–517 (1977).

Ganong, W.F.; Bernhard, W.F.; McMurrey, J.D.: The effect of hypothermia on the output of 17-hydroxycorticoids from the adrenal vein in the dog. Surgery, St. Louis 38: 506–512 (1955a).

Ganong, W.F.; Boryczka, A.T.; Lorenzen, L.C.; Egge, A.S.: Lack of effect of α-ethyltryptamine on ACTH secretion when blood pressure is held constant. Proc. Soc. exp. Biol. Med. 124: 558–559 (1967).

Ganong, W.F.; Gold, N.I.; Hume, D.M.: Effect of hypothalamic lesions on plasma

17 hydroxycorticoid response to immobilization in the dog. Fed. Proc. *14:* 54 (1955b).

Ganong, W.F.; Kramer, N.; Salmon, J.; Reid, I.A.; Lovinger, R.; Scapagnini, U.; Boryczka, A.T.; Shackelford, R.: Pharmacological evidence for inhibition of ACTH secretion by a central adrenergic system in the dog. Neuroscience *1:* 167–174 (1976).

Ganong, W.F.; Lieberman, A.H.; Daily, W.J.R.; Yuen, V.S.; Mulrow, P.J.; Luetscher, J.A., Jr.; Bailey, R.E.: Aldosterone secretion in dogs with hypothalamic lesions. Endocrinology *65:* 18–28 (1959).

Ganong, W.F.; Nolan, A.M.; Dowdy, A.; Luetscher, J.A.: The effect of hypothalamic lesions on adrenal secretion of cortisol, corticosterone, 11-desoxycortisol and aldosterone. Endocrinology *68:* 169–171 (1961).

Ganong, W.F.; Wise, B.L.; Shackleford, R.; Boryczka, A.T.; Zipf, B.: Site at which α-ethyltryptamine acts to inhibit the secretion of ACTH. Endocrinology *76:* 526–530 (1965).

Garcy, A.M.; Marotta, S.F.: Plasma cortisol of conscious cats during cerebroventricular perfusion with adrenergic, cholinergic and gabanergic antagonists. Neuroendocrinology *25:* 343–353 (1978).

Garris, D.R.: Diurnal fluctation of plasma cortisol levels in the guinea pig. Acta endocr., Copenh. *90:* 692–695 (1979).

Gaunt, R.; Renzi, A.A.; Chart, J.J.: Endocrine pharmacology of methyl reserpate derivatives. Endocrinology *71:* 527–535 (1962).

Gawel, M.J.; Park, D.M.; Alaghband-Zadeh, J.; Rose, F.C.: Exercise and hormonal secretion. Postgrad. med. J. *55:* 373–376 (1979).

George, R.; Way, E.L.: Adrenal cortical response of normal, adrenal demedullated, and hypophysectomized rats to morphine and methadone. J. Pharmac. exp. Ther. *113:* 23 (1955a).

George, R.; Way, E.L.: Studies on the mechanism of pituitary-adrenal activation by morphine. Br. J. Pharmacol. *10:* 260–264 (1955b).

George, R.; Way, E.L.: The hypothalamus as an intermediary for pituitary-adrenal activation by aspirin. J. Pharmac. exp. Ther. *119:* 310–316 (1957).

George, R.; Way, E.L.: The role of the hypothalamus in pituitary-adrenal activation and antidiuresis by morphine. J. Pharmac. exp. Ther. *125:* 111–115 (1959).

Georges, G.: Influence de la 5-hydroxytryptamine sur la réactivité du couple-hypophyso-surrénalien chez le rat. C. r. Séanc. Soc. Biol. *151:* 692–695 (1957).

Gerschman, R.; Fenn, W.O.: Ascorbic acid content of adrenal glands of rat in oxygen poisoning. Am. J. Physiol. *176:* 6–8 (1954).

Gershberg, H.; Long, C.N.H.: The activation of the adrenal cortex by insulin hypoglycemia. J.clin. Endocr. Metab. *8:* 587–588 (1948).

Ghosh, P.; Taneja, R.L.; Malhotra, S.C.; Kumar, V.; Ahuja, M.M.S.: A study of diurnal variation of adrenocortical function and recovery of the functional integrity of pituitary adrenal axis after long term corticosteroid therapy. Indian J. med. Res. *62:* 246–253 (1974a).

Ghosh, P.; Taneja, R.L.; Malhotra, S.C.; Kumar, V.; Ahuja, M.M.S.: A study on the circadian rhythm of plasma cortisol and urinary excretion of 17-hydroxycorticosteroids in relation to (a) changes in the environmental temperature, (b) glomerular filtration rate. Indian J. med. Res. *62:* 1392–1401 (1974b).

Gibbs, F.P.: Central nervous system lesions that block release of ACTH caused by traumatic stress. Am. J. Physiol. *217*: 78–83 (1969a).

Gibbs, F.P.: Area of pons necessary for traumatic stress-induced ACTH release under pentobarbital anesthesia. Am. J. Physiol. *217*: 84–88 (1969b).

Gibbs, F.P.: Correlation of plasma corticosterone levels with running activity in the blinded rat. Am. J. Physiol. *231*: 817–821 (1976).

Gibson, A.; Ginsburg, M.; Hall, M.; Hart, S.L.: The effects of opiate receptor agonists and antagonists on the stress-induced secretion of corticosterone in mice. Br. J. Pharmacol. *65*: 139–146 (1979).

Gibson, M.J.; Krieger, D.T.: Circadian corticosterone rhythm and stress response in rats with adrenal autotransplants. Am. J. Physiol. *240*: E363–E366 (1981).

Gillies, G.; Lowry, P.J.: Perfused rat isolated anterior pituitary cell column as bioassay for factor(s) controlling release of adrenocorticotropin: validation of a technique. Endocrinology *103*: 521–527 (1978).

Gillies, G.; Lowry, P.: Corticotrophin releasing factor may be modulated vasopressin. Nature, Lond. *278*: 463–464 (1979).

Gillies, G.; Lowry, P.J.: Corticotrophin releasing activity in extracts of the stalk median eminence of Brattleboro rats. J. Endocr. *84*: 65–73 (1980).

Gillies, G.; Van Wimersma Greidanus, T.B.; Lowry, P.J.: Characterization of rat stalk median eminence vasopressin and its involvement in adrenocorticotropin release. Endocrinology *103*: 528–534 (1978).

Giuliani, G.; Martini, L.; Pecile, A.: Midbrain section and release of ACTH following stress. Acta neuroveg. *23*: 21–34 (1961).

Givens, J.R.; Ney, R.L.; Nicholson, W.E.; Graber, A.L.; Liddle, G.W.: Absence of a normal diurnal variation of plasma ACTH in Cushing's disease. Clin. Res. *12*: 267 (1964).

Goldfien, A.; Ganong, W.F.: Adrenal medullary and adrenal cortical response to stimulation of diencephalon. Am. J. Physiol. *202*: 205–211 (1962).

Goldfien, A.; Zileli, M.S.; Despointes, R.H.; Bethune, J.E.: The effect of hypoglycemia on the adrenal secretion of epinephrine and norepinephrine in the dog. Endocrinology *62*: 749–757 (1958).

Gomez-Sanchez, C.; Holland, O.B.; Higgins, J.R.; Kem, D.C.; Kaplan, N.M.: Circadian rhythms of serum renin activity and serum corticosterone, prolactin, and aldosterone concentrations in the male rat on normal and low-sodium diets. Endocrinology *99*: 567–572 (1976).

Gonzalez-Luque, A.; L'Age, M.; Dhariwal, A.P.S.; Yates, F.E.: Stimulation of corticotropin release by corticotropin-releasing factor (CRF) or by vasopressin following intrapituitary infusions in unanesthetized dogs: inhibition of the responses by dexamethasone. Endocrinology *86*: 1134–1142 (1970).

Grahame-Smith, D.G.; Butcher, R.W.; Ney, R.L.; Sutherland, E.W.: Adenosine 3′,5′-monophosphate as the intracellular mediator of the action of adrenocorticotropic hormone on the adrenal cortex. J. biol. Chem. *242*: 5535–5541 (1967).

Gray, W.D.; Munson, P.L.: The rapidity of the adrenocorticotropic response of the pituitary to the intravenous administration of histamine. Endocrinology *48*: 471–481 (1951).

Greer, M.A.; Allen, C.F.; Gibbs, F.P.; Gullickson, C.: Pathways at the hypothalamic level through which traumatic stress activates ACTH secretion. Endocrinology *86*: 1404–1409 (1970).

Greer, M.A.; Rockie, C.: Inhibition by pentobarbital of ether-induced ACTH secretion in the rat. Endocrinology 83: 1247–1252 (1968).

Grim, C.; Winnacker, J.; Peters, T.; Gilbert, G.: Low renin, 'normal' aldosterone and hypertension: circadian rhythm of renin, aldosterone, cortisol and growth hormone. J. clin. Endocr. Metab. 39: 247–256 (1974).

Grimm, Y.; Kendall, J.W.: A study of feedback suppression of ACTH secretion utilizing glucocorticoid implants in the hypothalamus: the comparative effects of cortisol, corticosterone, and their 21-acetates. Neuroendocrinology 3: 55–63 (1968).

Grindeland, R.E.; Anderson, E.: Assay of vasopressin in blood. Fed. Proc. 23: 150 (1964).

Guignard, M.M.; Pesquies, P.C.; Serrurier, B.D.; Merino, D.B.; Reinberg, A.E.: Circadian rhythms in plasma levels of cortisol, dehydroepiandrosterone, Δ4-androstenedione, testosterone and dihydrotestosterone of healthy young men. Acta endocr., Copenh. 94: 536–545 (1980).

Guillemant, J.; Reinberg, A.; Guillemant, S.: Study of the role of adrenocortical cyclic AMP and cyclic GMP on ascorbic acid depletion and corticosteroidogenesis by analysis of circadian rhythms. Acta endocr., Copenh. 95: 382–387 (1980).

Guillemant, S.; Eurin, J.; Guillemant, J.; Reinberg, A.: Rythmes circadiens des nucléotides cycliques (AMP et GMP cycliques) du cortex surrénalien du rat mâle adulte. Annls Endocr. 39: 49–50 (1978).

Guillemin, R.: A re-evaluation of acetylcholine, adrenaline, nor-adrenaline and histamine as possible mediators of the pituitary adrenocorticotrophic activation by stress. Endocrinology 56: 248–255 (1955).

Guillemin, R.; Clayton, G.W.; Smith, J.D.; Lipscomb, H.S.: Measurement of free corticosteroids in rat plasma: physiological validation of a method. Endocrinology 63: 349–358 (1958).

Guillemin, R.; Dear, W.E.; Liebelt, R.A.: Nychthemeral variations in plasma free corticosteroid levels of the rat. Proc. Soc. exp. Biol. Med. 101: 394–395 (1959a).

Guillemin, R.; Dear, W.E.; Nichols, B., Jr.; Lipscomb, H.S.: ACTH releasing activity in vivo of a CRF preparation and lysine vasopressin. Proc. Soc. exp. Biol. Med. 101: 107–111 (1959b).

Gwinup, G.: Studies on the mechanism of vasopressin-induced steroid secretion in man. Metabolism 14: 1282–1286 (1965).

Haksar, A.; Péron, F.G.: Chlorpromazine: inhibition of ACTH and cyclic 3',5'-AMP stimulated corticosterone synthesis. Biochem. biophys. Res. Commun. 44: 1376–1380 (1971).

Haksar, A.; Péron, F.G.: Comparison of the Ca^{++} requirement for the steroidogenic effect of ACTH and dibutyryl cyclic AMP in rat adrenal cell suspensions. Biochem. biophys. Res. Commun. 47: 445–450 (1972).

Halász, B.; Pupp, L.: Hormone secretion of the anterior pituitary gland after physical interruption of all nervous pathways to the hypophysiotrophic area. Endocrinology 77: 553–562 (1965).

Halász, B.; Slusher, M.A.; Gorski, R.A.: Adrenocorticotrophic hormone secretion in rats after partial or total deafferentation of the medial basal hypothalamus. Neuroendocrinology 2: 43–55 (1967).

Halberg, F.; Albrecht, P.G.; Bittner, J.J.: Corticosterone rhythm of mouse adrenal in relation to serum corticosterone and sampling. Am. J. Physiol. 197: 1083–1085 (1959a).

Halberg, F.; Frank, G.; Harner, R.; Matthews, J.; Aaker, H.; Gravem, H.; Melby, J.: The adrenal cycle in men on different schedules of motor and mental activity. Experientia *17*: 282–284 (1961).

Halberg, F.; Peterson, R.E.; Silber, R.H.: Phase relations of 24-hour periodicities in blood corticosterone, mitoses in cortical adrenal parenchyma, and total body activity. Endocrinology *64*: 222–230 (1959b).

Hale, H.B.; Williams, E.W.; Anderson, J.E.; Ellis, J.P., Jr.: Endocrine and metabolic effects of short-duration hyperoxia. Aerospace Med. *35*: 449–451 (1964).

Hamanaka, Y.; Manabe, H.; Tanaka, H.; Monden, Y.; Uozumi, T.; Matsumoto, K.: Effects of surgery on plasma levels of cortisol, corticosterone and non-protein-bound-cortisol. Acta endocr., Copenh. *64*: 439–451 (1970).

Hamburger, C.: Substitution of hypophysectomy by the administration of chlorpromazine in the assay of corticotrophin. Acta endocr., Copenh. *20*: 383–390 (1955).

Hameed, J.M.A.; Haley, T.J.: Plasma and adrenal gland corticosterone levels after X-ray exposure in rats. Radiat. Res. *23*: 620–629 (1964).

Hammond, W.G.; Vandam, L.D.; Davis, J.M.; Carter, R.D.; Ball, M.R.; Moore, F.D.: Studies in surgical endocrinology. IV. Anesthetic agents as stimuli to change in corticosteroids and metabolism. Ann. Surg. *148*: 199–211 (1958).

Harwood, C.T.; Mason, J.W.: Effects of intravenous infusion of autonomic agents on peripheral blood 17-hydroxycorticosteroid levels in the dog. Am. J. Physiol. *186*: 445–452 (1956).

Harwood, C.T.; Mason, J.W.: Acute effects of tranquilizing drugs on the anterior pituitary-ACTH mechanism. Endocrinology *60*: 239–246 (1957).

Haus, E.; Lakatua, D.; Halberg, F.: The internal timing of several circadian rhythms in the blinded mouse. Expl Med. Surg. *25*: 7–45 (1967).

Haynes, R.C., Jr.: The activation of adrenal phosphorylase by the adrenocorticotropic hormone. J. biol. Chem. *233*: 1220–1222 (1958).

Haynes, R.C., Jr.; Berthet, L.: Studies on the mechanism of action of the adrenocorticotropic hormone. J. biol. Chem. *225*: 115–124 (1957).

Haynes, R.C., Jr.; Koritz, S.B.; Péron, F.G.: Influence of adenosine 3',5'-monophosphate on corticoid production by rat adrenal glands. J. biol. Chem. *234*: 1421–1423 (1959).

Hedge, G.A.: The effects of prostaglandins on ACTH secretion. Endocrinology *91*: 925–933 (1972).

Hedge, G.A.; De Wied, D.: Corticotropin and vasopressin secretion after hypothalamic implantation of atropine. Endocrinology *88*: 1257–1259 (1971).

Hedge, G.A.; Smelik, P.G.: Corticotropin release: inhibition by intrahypothalamic implantation of atropine. Science *159*: 891–892 (1968).

Hedge, G.A.; Smelik, P.G.: The action of dexamethasone and vasopressin on hypothalamic CRF production and release. Neuroendocrinology *4*: 242–253 (1969).

Hedge, G.A.; Yates, M.B.; Marcus, R.; Yates, F.E.: Site of action of vasopressin in causing corticotropin release. Endocrinology *79*: 328–340 (1966).

Hellman, L.; Nakada, F.; Curti, J.; Weitzman, E.D.; Kream, J.; Roffwarg, H.; Ellman, S.; Fukushima, D.K.; Gallagher, T.F.: Cortisol is secreted episodically by normal man. J. clin. Endocr. Metab. *30*: 411–422 (1970).

Hetzel, B.S.; Hine, D.C.: The effect of salicylates on the pituitary and suprarenal glands. Lancet *ii*: 94–97 (1951).

Heybach, J.P.; Vernikos-Danellis, J.: Inhibition of adrenocorticotrophin secretion during deprivation-induced eating and drinking in rats. Neuroendocrinology 28: 329–338 (1979).

Hilfenhaus, M.: Circadian rhythm of plasma renin activity, plasma aldosterone and plasma corticosterone in rats. Int. J. Chronobiol. 3: 213–229 (1976).

Hilfenhaus, M.; Herting, T.: The circadian rhythm of renal excretion in the rat: relationship between electrolyte and corticosteroid excretion. Contr. Nephrol., vol. 19, pp. 56–62 (Karger, Basel 1980).

Hill, C.D.; Singer, B.: Inhibition of the response to pituitary adrenocorticotrophic hormone in the hypophysectomized rat by circulatory corticosterone. J. Endocr. 42: 301–309 (1968).

Hillhouse, E.W.; Burden, J.; Jones, M.T.: The effect of various putative neurotransmitters on the release of corticotrophin releasing hormone from the hypothalamus of the rat in vitro. I. The effect of acetylcholine and noradrenaline. Neuroendocrinology 17: 1–11 (1975).

Hillhouse, E.W.; Jones, M.T.: Effect of bilateral adrenalectomy and corticosteroid therapy on the secretion of corticotrophin-releasing factor activity from the hypothalamus of the rat in vitro. J. Endocr. 71: 21–30 (1976).

Hilton, J.G.; Kruesi, O.R.; Nedeljkovic, R.I.; Scian, L.F.: Adreno-cortical and medullary responses to adenosine 3′, 5′ -monophosphate. Endocrinology 68: 908–913 (1961).

Hilton, J.G.; Scian, L.F.; Westermann, C.D.; Nakano, J.; Kruesi, O.R.: Vasopressin stimulation of the isolated adrenal glands: nature and mechanism of hydrocortisone secretion. Endocrinology 67: 298–310 (1960).

Hilton, J.G.; Weaver, D.C.; Muelheims, G.; Glaviano, V.V.; Wégria, R.: Perfusion of the isolated adrenals in situ. Am. J. Physiol. 192: 525–530 (1958).

Hirai, K.; Atkins, G.; Marotta, S.F.: 17-Hydroxycorticosteroid secretion during hypoxia in anesthetized dogs. Aerospace Med. 34: 814–816 (1963).

Hirai, M.; Nan, L.K.; Nakao, T.: Acute effect of morphine administration on secretion of adrenal corticosterone in rats. Endocr. jap. 17: 65–81 (1970).

Hirose, T.: Stimulation by prostaglandin E₂ of adrenal secretion of cortisol and corticosterone in infant dogs. IRCS med. Sci. 9: 909–910 (1981).

Hirose, T.; Matsumoto, I.; Aikawa, T.: Direct effect of histamine on cortisol and corticosterone production by isolated dog adrenal cells. J. Endocr. 76: 371–372 (1978).

Hirose, T.; Matsumoto, I.; Aikawa, T.: Effects of prostaglandin E₂ and dibutyryl cyclic AMP on the histamine-induced production of cortisol and corticosterone in isolated canine adrenal cells. J. Endocr. 82: 275–277 (1979).

Hirose, T.; Matsumoto, I.; Aikawa, T.; Suzuki, T.: Effect of histamine on the adrenal secretion of cortisol and corticosterone in hypophysectomized dogs. J. Endocr. 73: 539–540 (1977).

Hirose, T.; Matsumoto, I.; Suzuki, T.: Adrenal cortical secretory responses to histamine and cyanide in dogs with hypothalamic lesions. Neuroendocrinology 21: 304–311 (1976).

Hiroshige, T.: Role of vasopressin in the regulation of ACTH secretion: studies with intrapituitary injection technique. Med. J. Osaka Univ. 21: 161–180 (1971).

Hiroshige, T.; Abe, K.; Wada, S.; Kaneko, M.: Sex difference in circadian periodicity of CRF activity in the rat hypothalamus. Neuroendocrinology 11: 306–320 (1973).

Hiroshige, T.; Kunita, H.; Ogura, C.; Itoh, S.: Effects on ACTH release of intrapituitary injections of posterior pituitary hormones and several amines in the hypothalamus. Jap. J. Physiol. *18*: 609-619 (1968).

Hiroshige, T.; Sakakura, M.: Circadian rhythm of corticotropin-releasing activity in the hypothalamus of normal and adrenalectomized rats. Neuroendocrinology *7*: 25-36 (1971).

Hiroshige, T.; Sakakura, M.; Itoh, S.: Diurnal variation of corticotropin-releasing activity in the rat hypothalamus. Endocr. jap. *16*: 465-469 (1969a).

Hiroshige, T.; Sakakura, M.; Itoh, S.: Physiological significance of hypothalamic CRF in the regulation of ACTH secretion. Gunma Symp. Endocrinol. *6*: 231-247 (1969b).

Hiroshige, T.; Sato, T.: Circadian rhythm and stress-induced changes in hypothalamic content of corticotropin-releasing activity during postnatal development in the rat. Endocrinology *86*: 1184-1186 (1970a).

Hiroshige, T.; Sato, T.: Postnatal development of circadian rhythm of corticotropin-releasing activity in the rat hypothalamus. Endocr. jap. *17*: 1-6 (1970b).

Hiroshige, T.; Sato, T.; Abe, K.: Dynamic changes in the hypothalamic content of corticotropin-releasing factor following noxious stimuli: delayed response in early neonates in comparison with biphasic response in adult rats. Endocrinology *89*: 1287-1294 (1971).

Hiroshige, T.; Sato, T.; Ohta, R.; Itoh, S.: Increase of corticotropin-releasing activity in the rat hypothalamus following noxious stimuli. Jap. J. Physiol. *19*: 866-875 (1969c).

Hiroshige, T.; Wada-Okada, S.: Diurnal changes of hypothalamic content of corticotropin-releasing activity in female rats at various stages of the estrous cycle. Neuroendocrinology *12*: 316-319 (1973).

Hochman, A.; Bloch-Frankenthal, L.: The effect of low and high X-ray dosage on the ascorbic acid content of the suprarenal. Br. J. Radiol. *26*: 599-600 (1953).

Hodges, J.R.; Sadow, J.: Impairment of pituitary adrenocorticotrophic function by corticosterone in the blood. Br. J. Pharmacol. *30*: 385-391 (1967).

Hodges, J.R.; Vernikos, J.: Influence of circulating adrenocorticotrophin on the pituitary adrenocorticotrophic response to stress in the adrenalectomized rat. Nature, Lond. *182*: 725 (1958).

Hodges, J.R.; Vernikos, J.: Circulating corticotrophin in normal and adrenalectomized rats after stress. Acta endocr., Copenh. *30*: 188-196 (1959).

Hökfelt, B.: The effect of smoking on the production of adreno-cortical hormones. Acta med. scand. *170*: suppl. 369, pp. 123-124 (1961).

Hökfelt, B.; Hansson, B.-G.; Heding, L.G.; Nilsson, K.O.: Effect of insulin induced hypoglycaemia on the blood levels of catecholamines, glucagon, growth hormone, cortisol, C-peptide and proinsulin before and during medication with the cardioselective beta-receptor blocking agent metoprolol in man. Acta endocr., Copenh. *87*: 659-667 (1978).

Holbrook, M.M.; Dale, S.L.; Melby, J.C.: Peripheral plasma steroid concentrations in rats sacrificed by anoxia. J. Steroid Biochem. *13*: 1355-1358 (1980).

Holland, F.J.; Richards, G.E.; Kaplan, S.L.; Ganong, W.F.; Grumbach, M.M.: The role of biogenic amines in the regulation of growth hormone and corticotropin secretion in the trained conscious dog. Endocrinology *102*: 1452-1457 (1978).

Holmquest, D.L.; Retiene, K.; Lipscomb, H.S.: Circadian rhythms in rats: effects of random lighting. Science *152*: 662–664 (1966).

Holzbauer, M.: The corticosterone content of rat adrenals under different experimental conditions. J. Physiol., Lond. *139*: 294–305 (1957).

Holzbauer, M.; Vogt, M.: The action of chlorpromazine on diencephalic sympathetic activity and on the release of adrenocorticotrophic hormone. Br. J. Pharmacol. *9*: 402–407 (1954).

Honma, K.; Hiroshige, T.: Internal synchronization among several circadian rhythms in rats under constant light. Am. J. Physiol. *235*: R243–R249 (1978).

Honma, K.; Hiroshige, T.: Participation of brain catecholaminergic neurons in a self-sustained circadian oscillation of plasma corticosterone in the rat. Brain Res. *169*: 519–529 (1979).

Honn, K.V.; Chavin, W.: Prostaglandin modulation of the mechanism of ACTH action in the human adrenal. Biochem. biophys. Res. Commun. *73*: 164–170 (1976).

Honn, K.V.; Chavin, W.: Effects of A and B series prostaglandins on cAMP, cortisol and aldosterone production by the human adrenal. Biochem. biophys. Res. Commun. *76*: 977–982 (1977).

Hortling, H.; Puupponen, E.; Sundholm, I.: The vasopressin test as an aid in the evaluation of hypothalamo-pituitary-adrenal function. Acta med. scand. *189*: 479–484 (1971).

Hume, D.M.: The secretion of epinephrine, nor-epinephrine and corticosteroids in the adrenal venous blood of the dog following single and repeated trauma. Surg. Forum *8*: 111–115 (1957).

Hume, D.M.; Bell, C.C.; Bartter, F.: Direct measurement of adrenal secretion during operative trauma and convalescence. Surgery, St. Louis *52*: 174–187 (1962).

Hume, D.M.; Egdahl, R.H.: The importance of the brain in the endocrine response to injury. Ann. Surg. *150*: 697–712 (1959).

Hume, D.M.; Jackson, B.T.: Adrenal output of corticosteroids in the first 14 days after hypothalamic destruction in the dog. Fed. Proc. *18*: 481 (1959).

Hume, D.M.; Nelson, D.H.: Adrenal cortical function in surgical shock. Surg. Forum *5*: 568–575 (1955a).

Hume, D.M.; Nelson, D.H.: Effect of hypothalamic lesions on blood ACTH levels and 17-hydroxycorticosteroid secretion following trauma in the dog. J. clin. Endocr. Metab. *15*: 839–840 (1955b).

Ichikawa, S.; Sakamaki, T.; Tonooka, S.; Sugai, Y.: The diurnal rhythm of plasma aldosterone, plasma renin activity, plasma cortisol and serum growth hormone and subnormal responsiveness of aldosterone to angiotensin-II in the patients with normotensive acromegaly. Endocr. jap. *23*: 75–82 (1976).

Imura, H.; Matsukura, S.; Matsuyama, H.; Setsuda, T.; Miyake, T.: Adrenal steroidogenic effect of adenosine $3',5'$-monophosphate and its derivatives in vivo. Endocrinology *76*: 933–937 (1965).

Inoue, K.; Takahashi, K.; Takahashi, Y.: Influence of change in feeding regime and food deprivation on circadian rhythm of adrenal cortical activity in rats. Folia endocr. jap. *52*: 898–907 (1976).

Ishihara, I.; Komori, Y.; Maruyama, T.: Amygdala and adrenocortical response. Annu. Rep. Environ. Med. Nagoya Univ. *12*: 9–17 (1964).

Ishihara, I.; Komori, Y.; Maruyama, T.: Effect of corticosteroid administration on the adrenocorticotropic response to electrical stimulation of the hypothalamus in the dog. Annu. Rep. Environ. Med. Nagoya Univ. *14:* 19–30 (1965a).

Ishihara, I.; Komori, Y.; Tamura, Y.: Secretions of adrenal medulla and cortex after electrical stimulation of amygdala. Annu. Rep. Environ. Med. Nagoya Univ. *13:* 21–30 (1965b).

Itoh, S.: Role of vasopressin in the release of ACTH. Jap. J. Physiol. *7:* 213–221 (1957).

Itoh, S.; Arimura, A.: Effect of posterior pituitary hormone on the release of adrenocorticotrophic hormone. Nature, Lond. *174:* 37 (1954).

Itoh, S.; Katsuura, G.; Hirota, R.: Conditioned circadian rhythm of plasma corticosterone in the rat induced by food restriction. Jap. J. Physiol. *30:* 365–375 (1980).

Itoh, S.; Katsuura, G.; Hirota, R.; Botan, Y.: Circadian rhythm of plasma corticosterone in vagotomized rats. Experientia *37:* 380–381 (1981).

Itoh, S.; Nishimura, Y.; Yamamoto, M.; Takahashi, H.: Adrenocortical response to epinephrine in neurohypophysectomized rats. Jap. J. Physiol. *14:* 177–187 (1964).

Itoh, S.; Yamamoto, M.: Adrenocortical activity of adenohypophysectomized rats. Jap. J. Physiol. *14:* 265–269 (1964).

Jaanus, S.D.; Rosenstein, M.J.; Rubin, R.P.: On the mode of action of ACTH on the isolated perfused adrenal gland. J. Physiol., Lond. *209:* 539–556 (1970).

Jaanus, S.D.; Rubin, R.P.: The effect of ACTH on calcium distribution in the perfused cat adrenal gland. J. Physiol., Lond. *213:* 581–598 (1971).

Jacobs, J.J.: The effect of pinealectomy on adrenocortical rhythm in the blinded rat. Am. J. Anat. *139:* 437–442 (1974).

Jacoby, J.H.; Sassin, J.F.; Greenstein, M.; Weitzman, E.D.: Patterns of spontaneous cortisol and growth hormone secretion in rhesus monkeys during the sleep-waking cycle. Neuroendocrinology *14:* 165–173 (1974).

James, V.H.T.; Horner, M.W.; Moss, M.S.; Rippon, A.E.: Adrenocortical function in the horse. J. Endocr. *48:* 319–335 (1970).

Jarrett, D.B.; Newell, B.J.; Coghlan, J.P.; Scoggins, B.A.: Effect of adrenocorticotrophic hormone on the in vivo output of adenosine 3′5′ cyclic monophosphate and corticosteroids by the sheep adrenal gland. J. Steroid Biochem. *7:* 233–240 (1976).

Jenkins, J.S.: The pituitary-adrenal response to pyrogen; in James, Landon, The investigation of hypothalamic-pituitary-adrenal function. Mem. Soc. Endocrinol., vol. 17, pp. 205–212 (Cambridge University Press, London 1968).

Jensen, H.K.; Blichert-Toft, M.: Investigation of pituitary-adrenocortical function in the elderly during standardized operations and postoperative intravenous metyrapone test assessed by plasma cortisol, plasma compound S and eosinophil cell determinations. Acta endocr., Copenh. *67:* 495–507 (1971).

Jobin, M.; Ferland, L.; Labrie, F.: Effect of pharmacological blockade of ACTH and TSH secretion on the acute stimulation of prolactin release by exposure to cold and ether stress. Endocrinology *99:* 146–151 (1976).

Johnson, J.A.; Davis, J.O.; Brown, P.R.; Baumber, J.S.; Waid, R.A.: Evidence for a monophasic response in aldosterone and cortisol secretion to hemorrhage. Proc. Soc. exp. Biol. Med. *137:* 1121–1125 (1971).

Johnson, J.T.; Levine, S.: Influence of water deprivation on adrenocortical rhythms. Neuroendocrinology *11:* 268–273 (1973).

Johnston, S.D.; Mather, E.C.: Canine plasma cortisol (hydrocortisone) measured by radioimmunoassay: clinical absence of diurnal variation and results of ACTH stimulation and dexamethasone suppression tests. Am. J. vet. Res. *39:* 1766–1770 (1978).

Jones, M.T.; Brush, F.R.; Neame, R.L.B.: Characteristics of fast feedback control of cor-ticotrophin release by corticosteroids. J. Endocr. *55:*489–497 (1972).

Jones, M.T.; Hillhouse, E.W.; Burden, J.L.: Dynamics and mechanics of corticosteroid feedback at the hypothalamus and anterior pituitary gland. J. Endocr. *73:*405–417 (1977).

Jones, M.T.; Tiptaft, E.M.; Brush, F.R.; Fergusson, D.A.N.; Neame, R.L.B.: Evidence for dual corticosteroid-receptor mechanisms in the feedback control of adrenocor-ticotrophin secretion. J. Endocr. *60:*223–233 (1974).

Kägi, H.R.: Der Einfluss von Muskelarbeit auf die Blutkonzentration der Nebennieren-rindenhormone. Helv. med. Acta *3:*258–267 (1955).

Kakihana, R.; Blum, S.; Kessler, S.: Developmental study of pituitary-adrenocortical re-sponse in mice: plasma and brain corticosterone determination after histamine stress. J. Endocr. *60:*353–358 (1974).

Kakihana, R.; Butte, J.C.; Hathaway, A.; Noble, E.P.: Adrenocortical response to ethanol in mice: modification by chronic ethanol consumption. Acta endocr., Co-penh. *67:*653–664 (1971).

Kakihana, R.; Noble, E.P.; Butte, J.C.: Corticosterone response to ethanol in inbred strains of mice. Nature, Lond. *218:*360–361 (1968).

Kalant, H.; Hawkins, R.D.; Czaja, C.: Effect of acute alcohol intoxication on steroid output of rat adrenals in vitro. Am. J. Physiol. *204:*849–855 (1963).

Kaneko, M.; Hiroshige, T.: Fast, rate-sensitive corticosteroid negative feedback during stress. Am. J. Physiol. *234:*R39–R45 (1978a).

Kaneko, M.; Hiroshige, T.: Site of fast, rate-sensitive feedback inhibition of adrenocorti-cotropin secretion during stress. Am. J. Physiol. *234:*R46–R51 (1978b).

Kaneko, M.; Hiroshige, T.; Shinsako, J.; Dallman, M.F.: Diurnal changes in amplifica-tion of hormone rhythms in the adrenocortical system. Am. J. Physiol. *239:* R309–R316 (1980).

Kaplanski, J.; Dorst, W.; Smelik, P.G.: Pituitary-adrenal activity and depletion of brain catecholamines after α-methyl-*p*-tyrosine administration. Eur. J. Pharmacol. *20:* 238–240 (1972).

Kaplanski, J.; Nyakas, C.; Van Delft, A.M.L.; Smelik, P.G.: Effect of central early postnatal 6-hydroxydopamine administration on brain catecholamines and pituitary-adrenal function in adulthood. Neuroendocrinology *13:* 123–127 (1973/74).

Kaplanski, J.; Smelik, P.G.: Pituitary-adrenal activity and depletion of brain catechol-amines after central administration of 6-hydroxydopamine. Res. Commun. chem. Pathol. Pharmacol. *5:*263–271 (1973a).

Kaplanski, J.; Smelik, P.G.: Analysis of the inhibition of the ACTH release by hypothal-amic implants of atropine. Acta endocr., Copenh. *73:*651–659 (1973b).

Karaulova, L.K.: The effect of physical load of various duration on function of the hypo-thalamo-hypophyseal-adrenal system in rats. Fiziol. Zh. Sechenov. *59:* 1322–1325 (1973).

Kato, H.; Saito, M.; Suda, M.: Effect of starvation on the circadian adrenocortical rhythm in rats. Endocrinology *106:*918–921 (1980).

Katsuki, S.: A central control mechanism of the adrenocortical function: with special reference to the functional regulation within the hypothalamus. Acta neuroveg. *23:* 50–57 (1961).

Katsuki, S.; Ikemoto, T.; Shimada, H.; Hagiwara, F.; Kanai, J.: The functional relationship between the hypothalamus and the anterior pituitary-adrenal system. Endocr. jap. *2:* 303–312 (1955).

Katsuki, S.; Ito, M.; Watanabe, A.; Iino, K.; Yuji, S.; Kondo, S.: Effect of hypothalamic lesions on pituitary-adrenocortical responses to histamine and methopyrapone. Endocrinology *81:* 941–945 (1967).

Kawakami, M.; Seto, K.; Terasawa, E.; Yoshida, K.; Miyamoto, T.; Sekiguchi, M.; Hattori, Y.: Influence of electrical stimulation and lesion in limbic structure upon biosynthesis of adrenocorticoid in the rabbit. Neuroendocrinology *3:* 337–348 (1968a).

Kawakami, M.; Seto, K.; Yoshida, K.: Influence of corticosterone implantation in limbic structure upon biosynthesis of adrenocortical steroid. Neuroendocrinology *3:* 349–354 (1968b).

Kehlet, H.; Binder, C.; Engboek, C.: Imitation of the adreno-cortical response to surgery by intravenous infusion of synthetic human ACTH. Acta endocr., Copenh. *72:* 75–80 (1973).

Kendall, J.W.; Allen, C.: Studies on the glucocorticoid feedback control of ACTH secretion. Endocrinology *82:* 397–405 (1968).

Kendall, J.W.; Egans, M.L.; Stott, A.K.; Kramer, R.M.; Jacobs, J.J.: The importance of stimulus intensity and duration of steroid administration in suppression of stress-induced ACTH secretion. Endocrinology *90:* 525–530 (1972).

Kendall, J.W.; Grimm, Y.; Shimshak, G.: Relation of cerebrospinal fluid circulation to the ACTH-suppressing effects of corticosteroid implants in the rat brain. Endocrinology *85:* 200–208 (1969).

Kendall, J.W.; Roth, J.G.: Adrenocortical function in monkeys after forebrain removal or pituitary stalk section. Endocrinology *84:* 686–691 (1969).

Kendall, J.W., Jr.: Studies on inhibition of corticotropin and thyrotropin release utilizing microinjections into the pituitary. Endocrinology *71:* 452–455 (1962).

Kendall, J.W., Jr.; Allen, C.; Greer, M.A.: ACTH secretion in midbrain-transected rats. Endocrinology *77:* 1091–1096 (1965).

Kershbaum, A.; Pappajohn, D.J.; Bellet, S.; Hirabayashi, M.; Shafiiha, H.: Effect of smoking and nicotine on adrenocortical secretion. J. Am. med. Ass. *203:* 275–278 (1968).

Khazan, N.; Sulman, F.G.; Winnik, H.Z.: Activity of pituitary-adrenal cortex axis during acute and chronic reserpine treatment. Proc. Soc. exp. Biol. Med. *106:* 579–581 (1961).

Kim, C.; Kim, C.U.: Effect of partial hippocampal resection on stress mechanism in rats. Am. J. Physiol. *201:* 337–340 (1961).

Kimball, H.R.; Lipsett, M.B.; Odell, W.D.; Wolff, S.M.: Comparison of the effect of the pyrogens, etiocholanolone and bacterial endotoxin on plasma cortisol and growth hormone in man. J. clin. Endocr. Metab. *28:* 337–342 (1968).

Kimura, F.; Okano, H.; Kawakami, M.: Development of circadian rhythms in serum hormone levels in the immature female rat. Neuroendocrinology *32:* 19–23 (1981).

Kimura, M.: Influence of antidiuretic hormone on adrenal cortical activity in rats. Jap. J. Physiol. *4:* 24–31 (1954).

King, A.B.: The effect of exogenous dopamine on ACTH secretion. Proc. Soc. exp. Biol. Med. *130:*445–447 (1969).

King, A.B.; Thomas, J.A.: Effect of exogenous dopamine on rat adrenal ascorbic acid. J. Pharmac. exp. Ther. *159:*18–21 (1968).

Kitabchi, A.E.; Sharma, R.K.: Corticosteroidogenesis in isolated adrenal cells of rats. I. Effect of corticotropins and $3',5'$-cyclic nucleotides on corticosterone production. Endocrinology *88:*1109–1116 (1971).

Kitay, J.I.; Holub, D.A.; Jailer, J.W.: Inhibition of pituitary ACTH release: an extra-adrenal action of exogenous ACTH. Endocrinology *64:*475–482 (1959a).

Kitay, J.I.; Holub, D.A.; Jailer, J.W.: 'Inhibition' of pituitary ACTH release after administration of reserpine or epinephrine. Endocrinology *65:*548–554 (1959b).

Kloppenborg, P.W.C.; Island, D.P.; Liddle, G.W.; Michelakis, A.M.; Nicholson, W.E.: A method of preparing adrenal cell suspensions and its applicability to the in vitro study of adrenal metabolism. Endocrinology *82:*1053–1058 (1968).

Knapp, M.S.; Keane, P.M.; Wright, J.G.: Circadian rhythm of plasma 11-hydroxycorticosteroids in depressive illness, congestive heart failure, and Cushing's syndrome. Br. med. J. *ii:*27–30 (1967).

Knigge, K.M.: Adrenocortical response to stress in rats with lesions in hippocampus and amygdala. Proc. Soc. exp. Biol. Med. *108:*18–21 (1961).

Knigge, K.M.; Hays, M.: Evidence of inhibitive role of hippocampus in neural regulation of ACTH release. Proc. Soc. exp. Biol. Med. *114:*67–69 (1963).

Knigge, K.M.; Penrod, C.H.; Schindler, W.J.: In vitro and in vivo adrenal corticosteroid secretion following stress. Am. J. Physiol. *196:*579–582 (1959).

Knouff, R.A.; Brown, J.B.; Schneider, B.M.: Correlated chemical and histological studies of the adrenal lipids. I. The effect of extreme muscular activity on the adrenal lipids of the guinea pig. Anat. Rec. *79:*17–38 (1941).

Kobayashi, K.; Takahashi, K.: Prolonged food deprivation abolishes the circadian adrenocortical rhythm, but not the endogenous rhythm in rats. Neuroendocrinology *29:*207–214 (1979).

Kokka, N.; Eisenberg, R.M.; Garcia, J.; George, R.: Blood glucose, growth hormone, and cortisol levels after hypothalamic stimulation. Am. J. Physiol. *222:* 296–301 (1972).

Kolanowski, J.; Ortega, N.; Crabbé, J.: Réponse des cellules corticosurrénales isolées de rat et de cobaye à la corticotropine et à l'AMP cyclique. Annls Endocr. *35:*501–507 (1974).

Kosaka, K.: Studies on the mechanism of alloxan initial hyperglycemia. I. A role of pituitary and adrenal gland in connection with the effect of alloxan on the blood sugar. Tohoku J. exp. Med. *59:*379–390 (1954).

Kovács, K.; Dávid, M.A.; László, F.A.: Adrenocortical function in rats after lesion of the pituitary stalk. J. Endocr. *25:*9–18 (1962).

Kovács, K.; Horváth, E.; Kovács, B.M.; Kovács, G.S.; Petri, G.: The influence of chlorpromazine on the changes of the adrenal cortex caused by hypertonic saline in rats. Archs int. Pharmacodyn. Thér. *108:*170–179 (1956).

Kraicer, J.; Logothetopoulos, J.: Adrenal cortical response to insulin-induced hypoglycaemia in the rat. I. Adaptation to repeated daily injections of protamine zinc insulin. Acta endocr., Copenh. *44:*259–271 (1963a).

Kraicer, J.; Logothetopoulos, J.: Adrenal cortical response to insulin-induced hypoglycaemia in the rat. III. Lack of adaptation to repeated daily injections of a short acting insulin preparation. Acta endocr., Copenh. *44:*282–290 (1963b).

Kraicer, J.; Milligan, J.V.: Suppression of ACTH release from adenohypophysis by corticosterone: an in vitro study. Endocrinology *87:*371–376 (1970).

Kraicer, J.; Milligan, J.V.; Gosbee, J.L.; Conrad, R.G.; Branson, C.M.: In vitro release of ACTH: effects of potassium, calcium and corticosterone. Endocrinology *85:* 1144–1153 (1969).

Kraus, S.D.: Adrenal and plasma corticosterone and pituitary and plasma ACTH in alloxan diabetic rats. Proc. Soc. exp. Biol. Med. *143:*460–464 (1973).

Krey, L.C.; Lu, K.-H.; Butler, W.R.; Hotchkiss, J.; Piva, F.; Knobil, E.: Surgical disconnection of the medial basal hypothalamus and pituitary function in the rhesus monkey. II. GH and cortisol secretion. Endocrinology *96:* 1088–1093 (1975).

Krieger, D.T.: Diurnal pattern of plasma 17-hydroxycorticosteroids in pretectal and temporal lobe disease. J. clin. Endocr. Metab. *21:*695–698 (1961).

Krieger, D.T.: Effect of ocular enucleation and altered lighting regimens at various ages on the circadian periodicity of plasma corticosteroid levels in the rat. Endocrinology *93:*1077–1091 (1973).

Krieger, D.T.: Food and water restriction shifts corticosterone, temperature, activity and brain amine periodicity. Endocrinology *95:*1195–1201 (1974).

Krieger, D.T.; Allen, W.; Rizzo, F.; Krieger, H.P.: Characterization of the normal temporal pattern of plasma corticosteroid levels. J. clin. Endocr. Metab. *32:* 266–284 (1971).

Krieger, D.T.; Hauser, H.; Krey, L.C.: Suprachiasmatic nuclear lesions do not abolish food-shifted circadian adrenal and temperature rhythmicity. Science *197:* 398–399 (1977a).

Krieger, D.T.; Kreuzer, J.; Rizzo, F.A.: Constant light: effect on circadian pattern and phase reversal of steroid and electrolyte levels in man. J. clin. Endocr. Metab. *29:* 1634–1638 (1969).

Krieger, D.T.; Krieger, H.P.: Circadian pattern of plasma 17-hydroxycorticosteroid: alteration by anticholinergic agents. Science *155:* 1421–1422 (1967).

Krieger, D.T.; Krieger, H.P.: Effect of dexamethasone on pituitary-adrenal activation following intrahypothalamic implantation of 'neurotransmitters'. Endocrinology *87:* 179–182 (1970a).

Krieger, D.T.; Liotta, A.; Brownstein, M.J.: Corticotropin releasing factor distribution in normal and Brattleboro rat brain, and effect of deafferentation, hypophysectomy and steroid treatment in normal animals. Endocrinology *100:*227–237 (1977b).

Krieger, D.T.; Rizzo, F.: Serotonin mediation of circadian periodicity of plasma 17-hydroxycorticosteroids. Am. J. Physiol. *217:*1703–1707 (1969).

Krieger, D.T.; Silverberg, A.I.; Rizzo, F.; Krieger, H.P.: Abolition of circadian periodicity of plasma 17-OHCS levels in the cat. Am. J. Physiol. *215:*959–967 (1968).

Krieger, H.P.; Krieger, D.T.: Chemical stimulation of the brain: effect on adrenal corticoid release. Am. J. Physiol. *218:*1632–1641 (1970b).

Kuipers, F.; Ely, R.S.; Kelley, V.C.: Metabolism of steroids: the removal of exogenous 17-hydroxycorticosterone from the peripheral circulation in dogs. Endocrinology *62:*64–74 (1958).

Kuwamura, T.: Morphine and pituitary-adrenal function (effect in morphinized rats). Jikei med. J. *19:*79–94 (1972).

Kuwamura, T.; Nagao, S.; Nakaura, S.; Kawashima, K.; Tanaka, S.; Omori, Y.; Nakao, T.: Suppressive effect of dexamethasone on morphine-induced adrenocorticoido-genesis in rat. Endocrinol. jap. *20:* 359–364 (1973).

Kwaan, H.C.; Bartelstone, H.J.: Corticotropin release following injections of minute doses of arginine vasopressin into the third ventricle of the dog. Endocrinology *65:* 982–985 (1959).

L'Age, M.; Gonzalez-Luque, A.; Yates, F.E.: Adrenal blood flow dependence of cortisol secretion rate in unanesthetized dogs. Am. J. Physiol. *219:* 281–287 (1970).

Lammers, J.G.R.; De Wied, D.: The effect of chlorpromazine on exploratory and avoid-ance behaviour and on pituitary-adrenal activity in rats. Acta physiol. pharmac. neer. *12:* 169–170 (1963).

Lammers, J.G.R.; De Wied, D.: The blocking action of chlorpromazine on stress-in-duced pituitary-adrenal activity in nembutalized rats. Acta physiol. pharmac. neer. *13:* 103 (1964).

Landon, J.; James, V.H.T.; Stoker, D.J.: Plasma-cortisol response to lysine-vasopressin. Comparison with other tests of human pituitary-adrenocortical function. Lancet *ii:* 1156–1159 (1965).

Landon, J.; Wynn, V.; James, V.H.T.: The adrenocortical response to insulin-induced hypoglycaemia. J. Endocr. *27:* 183–192 (1963).

Langley, L.L.; Kilgore, W.G.: Carbon dioxide as a protecting and stressing agent. Am. J. Physiol. *180:* 277–278 (1955).

Lanier, L.P.; Van Hartesveldt, C.; Weis, B.J.; Isaacson, R.L.: Effects of differential hip-pocampal damage upon rhythmic and stress-induced corticosterone secretion in the rat. Neuroendocrinology *18:* 154–160 (1975).

Larsson, M.; Edqvist, L.-E.; Ekman, L.; Persson, S.: Plasma cortisol in the horse, diurnal rhythm and effects of exogenous ACTH. Acta vet. scand. *20:* 16–24 (1979).

Latner, A.L.; Cook, D.B.; Solanki, K.U.: Inhibition of binding of corticotropin-(1-24)-te-tracosapeptide (Synacthen) to membrane receptors of adrenal cortex by cortisol. Biochem. J. *164:* 477–480 (1977).

Lau, C.: Role of respiratory chemoreceptors in adrenocortical activation. Am. J. Physiol. *221:* 602–606 (1971a).

Lau, C.: Effects of O_2-CO_2 changes on hypothalamo-hypophyseal-adrenocortical acti-vation. Am. J. Physiol. *221:* 607–612 (1971b).

Lau, C.; Marotta, S.F.: Role of peripheral chemoreceptors on adrenocortical secretory rates during hypoxia. Aerospace Med. *40:* 1065–1068 (1969).

Laychock, S.G.; Rubin, R.P.: Indomethacin-induced alterations in corticosteroid and prostaglandin release by isolated adrenocortical cells of the cat. Br. J. Pharmacol. *57:* 273–278 (1976).

Leeman, S.E.; Glenister, D.W.; Yates, F.E.: Characterization of a calf hypothalamic ex-tract with adrenocorticotropin-releasing properties: evidence for a central nervous system site for corticosteroid inhibition of adrenocorticotropin release. Endocri-nology *70:* 249–262 (1962).

Lefkowitz, R.J.; Roth, J.; Pastan, I.: Effects of calcium on ACTH stimulation of the ad-renal: separation of hormone binding from adenyl cyclase activation. Nature, Lond. *228:* 864–866 (1970a).

Lefkowitz, R.J.; Roth, J.; Pricer, W.; Pastan, I.: ACTH receptors in the adrenal: specific binding of ACTH-^{125}I and its relation to adenyl cyclase. Proc. natn. Acad. Sci. USA 65:745–752 (1970b).

Legori, M.; Motta, M.; Zanisi, M.; Martini, L.: 'Short feedback loops' in ACTH control. Acta endocr., Copenh. suppl. 100:155 (1965).

Lehnert, G.; Leiber, H.; Schaller, K.H.: Plasmacortisol und Plasmacorticosteron im Anpassungsstadium der dosierten körperlichen Arbeit. Endokrinologie 52: 402–405 (1968).

Lengvári, I.; Branch, B.J.; Taylor, A.N.: The effect of perinatal thyroxine treatment on the development of the plasma corticosterone diurnal rhythm. Neuroendocrinology 24:65–73 (1977).

Lengvári, I.; Halász, B.: On the site of action of reserpine on ACTH secretion. J. neural Transm. 33:289–300 (1972).

Lengvári, I.; Halász, B.: Evidence for a diurnal fluctuation in plasma corticosterone levels after fornix transection in the rat. Neuroendocrinology 11:191–196 (1973).

Lengvári, I.; Liposits, Zs.: Diurnal changes in endogenous corticosterone content of some brain regions of rats. Brain Res. 124:571–575 (1977a).

Lengvári, I.; Liposits, Zs.: Return of diurnal plasma corticosterone rhythm long after frontal isolation of the medial basal hypothalamus in the rat. Neuroendocrinology 23:279–284 (1977b).

Lengvári, I.; Szelier, M.: Independence of diurnal rhythms of plasma corticosterone and hypothalamic serotonin content. Endokrinologie 76:297–302 (1980).

Leshner, A.I.; Toivola, P.T.K.; Terasawa, E.: Circadian variations in cortisol concentrations in the plasma of female rhesus monkeys. J. Endocr. 78:155–156 (1978).

Levin, R.; Fitzpatrick, K.M.; Levine, S.: Maternal influences on the ontogeny of basal levels of plasma corticosterone in the rat. Horm. Behav. 7:41–48 (1976).

Levin, R.; Levine, S.: Development of circadian periodicity in base and stress levels of corticosterone. Am. J. Physiol. 229:1397–1399 (1975).

Lewis, B.: A paper-chromatographic technique for the determination of plasma corticosteroids. J. clin. Path. 10:148–155 (1957).

Lewis, R.N.: Plasma hydrocortisone concentrations in relation to anaesthesia and surgery. Br. J. Anaesth. 35:84–90 (1963).

Librik, L.; Clayton, G.W.: The measurement of plasma nonesterified fatty acid levels following ACTH release in man. Metabolism 12:790–791 (1963).

Lightman, S.L.; James, V.H.T.; Linsell, C.; Mullen, P.E.; Peart, W.S.; Sever, P.S.: Studies of diurnal changes in plasma renin activity, and plasma noradrenaline, aldosterone and cortisol concentrations in man. Clin. Endocrinol. 14:213–223 (1981).

Linkola, J.; Fyhrquist, F.; Ylikahri, R.: Renin, aldosterone and cortisol during ethanol intoxication and hangover. Acta physiol. scand. 106:75–82 (1979).

Lippa, A.S.; Antelman, S.M.; Fahringer, E.E.; Redgate, E.S.: Relationship between catecholamines and ACTH: effects of 6-hydroxydopamine. Nature new Biol. 241: 24–25 (1973).

Long, C.N.H.: The relation of cholesterol and ascorbic acid to the secretion of the adrenal cortex. Recent Prog. Horm. Res. 1:99–122 (1947).

Long, C.N.H.; Fry, E.G.: Effect of epinephrine on adrenal cholesterol and ascorbic acid. Proc. Soc. exp. Biol. Med. 59:67–68 (1945).

Lorenzen, L.C.; Ganong, W.F.: Effect of drugs related to α-ethyltryptamine on stress-induced ACTH secretion in the dog. Endocrinology 80:889–892 (1967).

Lorenzo, L.; Mancheño, E.: Establecimiento del ritmo circadiano de cortisol en lactantes y niños de corta edad. An. Esp. Pediat. *12:*471–474 (1979).

Lotti, V.J.; Kokka, N.; George, R.: Pituitary-adrenal activation following intrahypothalamic microinjection of morphine. Neuroendocrinology *4:*326–332 (1969).

Louis, T.M.; Challis, J.R.G.; Robinson, J.S.; Thorburn, G.D.: Rapid increase of foetal corticosteroids after prostaglandin E2. Nature, Lond. *264:*797–799 (1976).

Lowry, P.J.; McMartin, C.; Peters, J.: Properties of a simplified bioassay for adrenocorticotrophic activity using the steroidogenic response of isolated adrenal cells. J. Endocr. *59:*43–55 (1973).

Lutz-Bucher, B.; Koch, B.; Mialhe, C.: Comparative in vitro studies on corticotropin releasing activity of vasopressin and hypothalamic median eminence extract. Neuroendocrinology *23:*181–192 (1977).

Lutz-Bucher, B.; Koch, B.; Mialhe, C.; Briaud, B.: Involvement of vasopressin in corticotropin-releasing effect of hypothalamic median eminence extract. Neuroendocrinology *30:*178–182 (1980).

Lymangrover, J.R.: Adrenocorticotrophic hormone and cyclic adenosine monophosphate effect on mouse adrenal cortical cell membrane potential. Experientia *36:* 613–614 (1980).

Lymangrover, J.R.; Matthews, E.K.; Saffran, M.: Membrane potential changes of mouse adrenal zona fasciculata cells in response to adrenocorticotropin and adenosine 3′,5′-monophosphate. Endocrinology *110:*462–468 (1982).

Mackie, C.; Schulster, D.: Phosphodiesterase activity and the potentiation by theophylline of adrenocorticotrophin stimulated steroidogenesis and adenosine 3′,5′-monophosphate levels in isolated rat adrenal cells. Biochem. biophys. Res. Commun. *53:*545–551 (1973).

Mahfouz, M.; Ezz, E.A.: The effect of reserpine and chlorpromazine on the response of the rat to acute stress. J. Pharmac. exp. Ther. *123:*39–42 (1958).

Maickel, R.P.; Westermann, E.O.; Brodie, B.B.: Effects of reserpine and cold-exposure on pituitary-adrenocortical function in rats. J. Pharmac. exp. Ther. *134:* 167–175 (1961).

Makara, G.B.; Stark, E.: Effect of gamma-aminobutyric acid (GABA) and GABA antagonist drugs on ACTH release. Neuroendocrinology *16:*178–190 (1974).

Makara, G.B.; Stark, E.: The effects of cholinomimetic drugs and atropine on ACTH release. Neuroendocrinology *21:*31–41 (1976).

Makara, G.B.; Stark, E.; Marton, J.; Mészáros, T.: Corticotrophin release induced by surgical trauma after transection of various afferent nervous pathways to the hypothalamus. J. Endocr. *53:*389–395 (1972).

Makara, G.B.; Stark, E.; Palkovits, M.: Afferent pathways of stressfull stimuli: corticotrophin release after hypothalamic deafferentation. J. Endocr. *47:*411–416 (1970).

Makara, G.B.; Stark, E.; Palkovits, M.: Reevaluation of the pituitary-adrenal response to ether in rats with various cuts around the medial basal hypothalamus. Neuroendocrinology *30:*38–44 (1980).

Makara, G.B.; Stark, E.; Palkovits, M.; Révész, T.; Mihály, K.: Afferent pathways of stressful stimuli: corticotrophin release after partial deafferentation of the medial basal hypothalamus. J. Endocr. *44:*187–193 (1969).

Malayan, S.A.; Ramsay, D.J.; Keil, L.C.; Reid, I.A.: Effects of increases in plasma vasopressin concentration on plasma renin activity, blood pressure, heart rate, and plasma corticosteroid concentration in conscious dogs. Endocrinology *107:* 1899–1904 (1980).

Mandell, A.J.; Chapman, L.F.; Rand, R.W.; Walter, R.D.: Plasma corticosteroids: changes in concentration after stimulation of hippocampus and amygdala. Science *139:* 1212 (1963).

Mangili, G.; Motta, M.; Muciaccia, W.; Martini, L.: Midbrain stress and ACTH secretion. Eur. Rev. Endocrinol. *1:* 247–253 (1965).

Mann, D.R.; Cost, M.G.; Jacobson, C.D.; Macfarland, L.A.: Adrenal gland rhythmicity and pituitary regulation of adrenal steroid secretion. Proc. Soc. exp. Biol. Med. *156:* 441–445 (1977).

Marantz, R.; Sachar, E.J.; Weitzman, E.; Sassin, J.: Cortisol and GH responses to *D*- and *L*-amphetamine in monkeys. Endocrinology *99:* 459–465 (1976).

Maren, T.H.: Pharmacology of nicotine: antipyretic, renal and adrenocorticotrophic effects. Proc. Soc. exp. Biol. Med. *77:* 521–523 (1951).

Marotta, S.F.: Roles of aortic and carotid chemoreceptors in activating the hypothalamo-hypophyseal-adrenocortical system during hypoxia. Proc. Soc. exp. Biol. Med. *141:* 915–922 (1972a).

Marotta, S.F.: Comparative effects of hypoxia, adrenocorticotropin and methylcholine on adrenocortical secretory rates. Proc. Soc. exp. Biol. Med. *141:* 923–927 (1972b).

Marotta, S.F.; Hirai, K.; Atkins, G.: Secretion of 17-hydroxycorticosteroids in conscious and anesthetized dogs exposed to simulated altitude. Proc. Soc. exp. Biol. Med. *114:* 403–405 (1963).

Marotta, S.F.; Hirai, K.; Atkins, G.: Adrenocortical secretion in anesthetized dogs during hyperoxia, hypoxia and positive pressure breathing. Proc. Soc. exp. Biol. Med. *118:* 922–926 (1965).

Marotta, S.F.; Lanuza, D.M.; Hiles, L.G.: Diurnal variations in plasma corticosterone and cations of male rats on two lighting schedules. Hormone metabol. Res. *6:* 329–331 (1974).

Marotta, S.F.; Malasanos, L.J.; Boonayathap, U.: Inhibition of the adrenocortical response to hypoxia by dexamethasone. Aerospace Med. *44:* 1–4 (1973).

Marotta, S.F.; Sithichoke, N.; Garcy, A.M.; Yu, M.: Adrenocortical responses of rats to acute hypoxic and hypercapnic stresses after treatment with aminergic agents. Neuroendocrinology *20:* 182–192 (1976).

Martel, R.R.; Westermann, E.O.; Maickel, R.P.: Dissociation of reserpine-induced sedation and ACTH hypersecretion. Life Sci. *4:* 151–155 (1962).

Martin, J.; Endröczi, E.; Bata, G.: Effect of the removal of amygdalic nuclei on the secretion of adrenal cortical hormones. Acta physiol. hung. *14:* 131–134 (1958).

Martin, M.M.; Martin, A.L.A.: Simultaneous fluorometric determination of cortisol and corticosterone in human plasma. J. clin. Endocr. Metab. *28:* 137–145 (1968).

Martini, L.; Pecile, A.; Saito, S.; Tani, F.: The effect of midbrain transection on ACTH release. Endocrinology *66:* 501–507 (1960).

Mason, J.W.: Plasma 17-hydroxycorticosteroid response to hypothalamic stimulation in the conscious rhesus monkey. Endocrinology *63:* 403–411 (1958).

Mason, J.W.: Plasma 17-hydroxycorticosteroid levels during electrical stimulation of the amygdaloid complex in conscious monkeys. Am. J. Physiol. *196:* 44–48 (1959).

Mason, J.W.; Harwood, C.T.; Rosenthal, N.R.: Influence of some environmental factors on plasma and urinary 17-hydroxycorticosteroid levels in the rhesus monkey. Am. J. Physiol. *190:*429–433 (1957).

Matheson, G.K.; Branch, B.J.; Taylor, A.N.: Effects of amygdaloid stimulation on pituitary-adrenal activity in conscious cats. Brain Res. *32:*151–167 (1971).

Matsuda, K.; Duyck, C.; Kendall, J.W., Jr.; Greer, M.A.: Pathways by which traumatic stress and ether induce increased ACTH release in the rat. Endocrinology *74:* 981–985 (1964).

Matsuda, K.; Kendall, J.W., Jr.; Duyck, C.; Greer, M.A.: Neural control of ACTH secretion: effect of acute decerebration in the rat. Endocrinology *72:*845–852 (1963).

Matsui, N.; Plager, J.E.: Rate of blood glucose fall as a determinant factor in insulin-induced adrenocortical stimulation. Endocrinology *79:*737–744 (1966).

Matsumoto, I.; Hirose, T.; Aikawa, T.: Direct effect of histamine on the production of adrenocortical hormone by guinea-pig adrenal cells. Jap. J. Physiol. *31:* 605–608 (1981).

Matsuoka, H.; Tan, S.Y.; Mulrow, P.J.: Effects of prostaglandins on adrenal steroidogenesis in the rat. Prostaglandins *19:*291–298 (1980).

Matthews, E.K.: Membrane potential measurement in cells of the adrenal gland. J. Physiol., Lond. *189:*139–148 (1967).

Matthews, E.K.; Saffran, M.: Steroid production and membrane potential measurement in cells of the adrenal cortex. J. Physiol., Lond. *189:*149–161 (1967).

Matthews, E.K.; Saffran, M.: Effect of ACTH on the electrical properties of adrenocortical cells. Nature, Lond. *219:*1369–1370 (1968).

Matthews, E.K.; Saffran, M.: Ionic dependence of adrenal steroidogenesis and ACTH-induced changes in the membrane potential of adrenocortical cells. J. Physiol., Lond. *234:*43–64 (1973).

McCann, S.M.: Effect of hypothalamic lesions on the adrenal cortical response to stress in the rat. Am. J. Physiol. *175:*13–20 (1953).

McCann, S.M.: The ACTH-releasing activity of extracts of the posterior lobe of the pituitary in vivo. Endocrinology *60:*664–676 (1957).

McCann, S.M.; Antunes-Rodrigues, J.; Nallar, R.; Valtin, H.: Pituitary-adrenal function in the absence of vasopressin. Endocrinology *79:*1058–1064 (1966).

McCann, S.M.; Brobeck, J.R.: Evidence for a role of the supraopticohypophyseal system in regulation of adrenocorticotrophin secretion. Proc. Soc. exp. Biol. Med. *87:* 318–324 (1954).

McCann, S.M.; Haberland, P.: Further studies on the regulation of pituitary ACTH in rats with hypothalamic lesions. Endocrinology *66:*217–221 (1960).

McCarthy, J.L.; Corley, R.C.; Zarrow, M.X.: Diurnal rhythm in plasma corticosterone and lack of diurnal rhythm in plasma compound F-like material in the rat. Proc. Soc. exp. Biol. Med. *104:*787–789 (1960).

McClure, D.J.: The diurnal variation of plasma cortisol levels in depression. J. psychosom. Res. *10:*189–195 (1966).

McDonald, I.R.; Goding, J.R.; Wright, R.D.: Transplantation of the adrenal gland of the sheep to provide access to its blood supply. Aust. J. exp. Biol. med. Sci. *36:* 83–96 (1958).

McDonald, R.K.; Weise, V.K.: Effect of arginine-vasopressin and lysine-vasopressin on plasma 17-hydroxycorticosteroid levels in man. Proc. Soc. exp. Biol. Med. *92:* 481–483 (1956).

McDonald, R.K.; Weise, V.K.; Peterson, R.E.: Effect of aspirin and reserpine on adreno-cortical response to Piromen in man. Proc. Soc. exp. Biol. Med. *93:*343–348 (1956).

McHugh, P.R.; Black, W.C.; Mason, J.W.: Some hormonal responses to electrical self-stimulation in the *Macaca mulatta.* Am. J. Physiol. *210:*109–113 (1966).

McHugh, P.R.; Smith, G.P.: Plasma 17-OHCS response to amygdaloid stimulation with and without afterdischarges. Am. J. Physiol. *212:*619–622 (1967).

McKinney, W.T., Jr.; Prange, A.J., Jr.; Majchowicz, E.; Schlesinger, K.: Plasma cortico-sterone changes following alterations in brain norepinephrine and serotonin. Dis. nerv. Syst. *32:*308–313 (1971).

McMurtry, J.P.; Wexler, B.C.: Hypersensitivity of spontaneously hypertensive rats (SHR) to heat, ether, and immobilization. Endocrinology *108:*1730–1736 (1981).

McNatty, K.P.; Cashmore, M.; Young, A.: Diurnal variation in plasma cortisol levels in sheep. J. Endocr. *54:*361–362 (1972).

McNatty, K.P.; Thurley, D.C.: The episodic nature of changes in ovine plasma cortisol levels and their response to adrenaline during adaptation to a new environment. J. Endocr. *59:*171–180 (1973).

Melby, J.C.; Egdahl, R.H.; Spink, W.W.: Effect of Brucella endotoxin on adrenocortical function in the dog. Fed. Proc. *16:*425–426 (1957).

Melby, J.C.; Egdahl, R.H.; Spink, W.W.: Secretion and metabolism of cortisol after in-jection of endotoxin. J. Lab. clin. Med. *56:*50–62 (1960).

Michael, R.P.; Setchell, K.D.R.; Plant, T.M.: Diurnal changes in plasma testosterone and studies on plasma corticosteroids in non-anaesthetized male rhesus monkeys (*Ma-caca mulatta*). J. Endocr. *63:*325–335 (1974).

Migeon, C.J.; French, A.B.; Samuels, L.T.; Bowers, J.Z.: Plasma 17-hydroxycorticoste-roid levels and leucocyte values in the rhesus monkey, including normal variation and the effect of ACTH. Am. J. Physiol. *182:*462–468 (1955).

Migeon, C.J.; Tyler, F.H.; Mahoney, J.P.; Florentin, A.A.; Castle, H.; Bliss, E.L.; Samu-els, L.T.: The diurnal variation of plasma levels and urinary excretion of 17-hy-droxycorticosteroids in normal subjects, night workers and blind subjects. J. clin. Endocr. Metab. *16:*622–633 (1956).

Milcu, S.M.; Bogdan, C.; Nicolau, G.Y.; Cristea, A.: Cortisol circadian rhythm in 70–100-year-old subjects. Revue roum. Méd.-Endocrinol. *16:*29–39 (1978).

Miller, M.; Moses, A.M.: Effect of temperature and dexamethasone on the plasma 17-hy-droxycorticoid and growth hormone responses to pyrogen. J. clin. Endocr. Metab. *28:*1056–1063 (1968).

Mitamura, T.: Effect of anesthetics on the pituitary-adrenocortical system of the rat. Acta med. nagasaki. *5:*58–62 (1960a).

Mitamura, T.: Effect of sodium pentobarbital anesthesia upon adrenal ascorbic acid de-pletion induced by morphine, histamine, adrenaline and insulin. Acta med. naga-saki. *5:*63–66 (1960b).

Miyabo, S.; Hisada, T.: Sex difference in ontogenesis of circadian adrenocortical rhythm in cortisone-primed rats. Nature, Lond. *256:*590–592 (1975).

Miyabo, S.; Ooya, E.; Hayashi, S.: Effect of intrahypothalamic implantation of cortisone acetate on the onset of circadian corticosterone rhythm in neonatal female rats. Neuroendocrinology *33:*47–51 (1981).

Miyatake, A.; Morimoto, Y.; Oishi, T.; Hanasaki, N.; Sugita, Y.; Iijima, S.; Teshima, Y.; Hishikawa, Y.; Yamamura, Y.: Circadian rhythm of serum testosterone and its re-

lation to sleep: comparison with the variation in serum luteinizing hormone, prolactin, and cortisol in normal men. J. clin. Endocr. Metab. *51:* 1365–1371 (1980).

Moberg, G.P.: Site of action of endotoxins on hypothalamic-pituitary-adrenal axis. Am.
J. Physiol. *220:* 397–400 (1971).

Moberg, G.P.; Bellinger, L.L.; Mendel, V.E.: Effect of meal feeding on daily rhythms of
plasma corticosterone and growth hormone in the rat. Neuroendocrinology *19:*
160–169 (1975).

Moberg, G.P.; Scapagnini, U.; De Groot, J.; Ganong, W.F.: Effect of sectioning the fornix on diurnal fluctuation in plasma corticosterone levels in the rat. Neuroendocrinology *7:* 11–15 (1971).

Montanari, R.; Stockham, M.A.: Effects of single and repeated doses of reserpine on the
secretion of adrenocorticotrophic hormone. Br. J. Pharmacol. *18:* 337–345 (1962).

Moore, R.Y.; Eichler, V.B.: Loss of a circadian adrenal corticosterone rhythm following
suprachiasmatic lesions in the rat. Brain Res. *42:* 201–206 (1972).

Morimoto, Y.; Arisue, K.; Yamamura, Y.: Relationship between circadian rhythm
of food intake and that of plasma corticosterone and effect of food restriction
on circadian adrenocortical rhythm in the rat. Neuroendocrinology *23:* 212–222
(1977).

Morimoto, Y.; Oishi, T.; Arisue, K.; Ogawa, Z.; Tanaka, F.; Yano, S.; Yamamura, Y.:
Circadian rhythm of plasma corticosteroid in adult female rats: chronological shifts
in abnormal lighting regimens and connection with oestrous cycle. Acta endocr.,
Copenh. *80:* 527–541 (1975).

Morimoto, Y.; Oishi, T.; Arisue, K.; Yamamura, Y.: Effect of food restriction and its
withdrawal on the circadian adrenocortical rhythm in rats under constant dark or
constant lighting condition. Neuroendocrinology *29:* 77–83 (1979).

Morimoto, Y.; Oishi, T.; Hanasaki, N.; Miyatake, A.; Noma, K.; Yamamura, Y.: Rela-
tive potency in acute and chronic suppressive effects of prednisolone and betamethasone on the hypothalamic-pituitary-adrenal axis in man. Endocr. jap. *27:*
659–666 (1980).

Motta, M.; Mangili, G.; Martini, L.: A 'short' feedback loop in the control of ACTH secretion. Endocrinology *77:* 392–395 (1965).

Motta, M.; Sterescu, N.; Piva, F.; Martini, L.: The participation of 'short' feedback
mechanisms in the control of ACTH and TSH secretion. Acta neurol. belg. *69:*
501–507 (1969).

Moussatché, H.; Pereira, N.A.: Release of adrenocorticotrophin by 5-hydroxytrypt-
amine. Acta physiol. lat. am. *7:* 71–75 (1957).

Moyle, W.R.; Kong, Y.C.; Ramachandran, J.: Steroidogenesis and cyclic adenosine
3′,5′-monophosphate accumulation in rat adrenal cells. J. biol. Chem. *248:* 2409–
2417 (1973).

Muelheims, G.H.; Francis, F.E.; Kinsella, R.A., Jr.: Suppression of the hypothalamic-pituitary-adrenal axis in the newborn dog. Endocrinology *85:* 365–367 (1969).

Mulder, G.H.; Smelik, P.G.: A superfusion system technique for the study of the sites of
action of glucocorticoids in the rat hypothalamus-pituitary-adrenal system in vitro.
I. Pituitary cell superfusion. Endocrinology *100:* 1143–1152 (1977).

Müller-Hess, R.; Geser, C.A.; Jéquier, E.; Felber, J.-P.; Vannotti, A.: Effects of adrenaline on insulin-induced release of growth hormone and cortisol in man. Acta endocr., Copenh. *75:* 260–273 (1974).

Mulrow, P.J.; Ganong, W.F.: The effect of hemorrhage upon aldosterone secretion in normal and hypophysectomized dogs. J. clin. Invest. *40:* 579–585 (1961).

Munson, P.L.; Briggs, F.N.: The mechanism of stimulation of ACTH secretion. Recent Prog. Horm. Res. *11:* 83–117 (1955).

Myers, J.L.; O'Hara, E.T.; Heizer, J.W.; Liberman, H.; Marks, L.J.: The effect of the acute administration of corticosteroids on the adrenocortical response to surgery. Surg. Forum *11:* 131–134 (1960).

Nagareda, C.S.; Gaunt, R.: Functional relationship between the adrenal cortex and posterior pituitary. Endocrinology *48:* 560–567 (1951).

Nagata, G.: Influences of chlorpromazine upon the pituitary function – especially ACTH and gonadotrophin secretion. Annu. Rep. Shionogi Res. Lab. *7:* 71–86 (1957).

Nakasone, K.: Influence of X-irradiation of the adrenal gland on corticoid secretory activity in dogs. Tohoku J. exp. Med. *106:* 83–87 (1972).

Nakasone, K.; Shimizu, T.: Blockade of hypothalamo-hypophyseal-adrenocortical mechanism caused by dexamethasone. Acta med. nagasaki. *15:* 49–57 (1971).

Narita, S.: Comparative studies on the adrenal medullary and cortical response to histamine. Tohoku J. exp. Med. *104:* 349–357 (1971).

Nasmyth, P.A.: The effect of some sympathomimetic amines on the ascorbic acid content of rats' adrenal glands. J. Physiol., Lond. *110:* 294–300 (1950).

Nasmyth, P.A.: The effects of histamine and antihistamines on the ascorbic acid content of rat's adrenal glands. J. Physiol., Lond. *112:* 215–222 (1951).

Nasmyth, P.A.: Factors influencing the effect of morphine sulphate on the ascorbic acid content of rats' adrenal glands. Br. J. Pharmacol. *9:* 95–99 (1954).

Nasmyth, P.A.: The effect of chlorpromazine on adrenocortical activity in stress. Br. J. Pharmacol. *10:* 336–339 (1955).

Natelson, B.H.; Smith, G.P.; Stokes, P.E.; Root, A.W.: Plasma 17-hydroxycorticosteroids and growth hormone during defense reactions. Am. J. Physiol. *226:* 560–568 (1974).

Nathan, R.S.; Sachar, E.J.; Langer, G.; Tabrizi, M.A.; Halpern, F.S.: Diurnal variation in the response of plasma prolactin, cortisol, and growth hormone to insulin-induced hypoglycemia in normal men. J. clin. Endocr. Metab. *49:* 231–235 (1979).

Naumenko, E.V.: Role of adrenergic and cholinergic structures in the control of the pituitary-adrenal system. Endocrinology *80:* 69–76 (1967).

Naumenko, E.V.: Hypothalamic chemoreactive structures and the regulation of pituitary-adrenal function. Effects of local injections of norepinephrine, carbachol and serotonin into the brain of guinea pigs with intact brains and after mesencephalic transection. Brain Res. *11:* 1–10 (1968).

Nazar, K.: Influence of muscular work on the plasma levels of 17-hydroxycorticosteroids. Acta physiol. pol. *16:* 195–206 (1965).

Nazar, K.: Relation between total amount of muscular work performed and changes in 17-hydroxysteroids level in the blood. Acta physiol. pol. *17:* 767–773 (1966).

Nelson, D.H.; Egdahl, R.H.; Hume, D.M.: Corticosteroid secretion in the adrenal vein of the non-stressed dog exposed to cold. Endocrinology *58:* 309–314 (1956).

Nelson, D.H.; Samuels, L.T.: A method for the determination of 17-hydroxycorticosteroids in blood: 17-hydroxycorticosterone in the peripheral circulation. J. clin. Endocr. Metab. *12:* 519–526 (1952).

Newman, A.E.; Redgate, E.S.; Farrell, G.: The effects of diencephalic-mesencephalic lesions on aldosterone and hydrocortisone secretion. Endocrinology *63:* 723–736 (1958).

Ney, R.L.; Shimizu, N.; Nicholson, W.E.; Island, D.P.; Liddle, G.W.: Correlation of plasma ACTH concentration with adrenocortical response in normal human subjects, surgical patients, and patients with Cushing's disease. J. clin. Invest. *42:* 1669–1677 (1963).

Nichols, B., Jr.; Guillemin, R.: Endogenous and exogenous vasopressin on ACTH release. Endocrinology *64:* 914–920 (1959).

Nichols, T.; Nugent, C.A.; Tyler, F.H.: Diurnal variation in suppression of adrenal function by glucocorticoids. J. clin. Endocr. Metab. *25:* 343–349 (1965).

Nikodijevic, O.; Maickel, R.P.: Some effects of morphine on pituitary-adrenocortical function in the rat. Biochem. Pharmac. *16:* 2137–2142 (1967).

Nilsson, E.; Arner, B.; Hedner, P.: Corticosteroid concentration in plasma during anaesthesia and at operation. Acta chir. scand. *126:* 281–288 (1963).

Nims, L.F.; Sutton, E.: Adrenal cholesterol, liver glycogen and water consumption of fasting and X-irradiated rats. Am. J. Physiol. *177:* 51–54 (1954).

Nolten, W.E.; Lindheimer, M.D.; Rueckert, P.A.; Oparil, S.; Ehrlich, E.N.: Diurnal patterns and regulation of cortisol secretion in pregnancy. J. clin. Endocr. Metab. *51:* 466–472 (1980).

North, N.; Nims, L.F.: Time-dose study of biochemical responses of rats to X-radiation. Fed. Proc. *8:* 119–120 (1949).

Notter, G.; Gemzell, C.A.: Veränderungen im Plasmaspiegel der 17-Hydroxy-Corticosteroide und Anzahl der eosinophilen Leukozyten beim Menschen nach Röntgenbestrahlung der Nebennieren. Strahlentherapie *99:* 203–212 (1956).

Nowell, N.W.: Studies in the activation and inhibition of adrenocorticotrophin secretion. Endocrinology *64:* 191–201 (1959).

Nugent, C.A.; Eik-Nes, K.; Kent, H.S.; Samuels, L.T.; Tyler, F.H.: A possible explanation for Cushing's syndrome associated with adrenal hyperplasia. J. clin. Endocr. Metab. *20:* 1259–1268 (1960).

Obled, C.; Arnal, M.; Grizard, J.: Influence du comportement alimentaire sur le rythme circadien de la corticostérone plasmatique chez le rat en croissance. C. r. hebd. Séanc. Acad. Sci., Paris sér. D *284:* 195–198 (1977).

Ohler, E.A.; Sevy, R.W.: Inhibition of stress induced adrenal ascorbic acid depletion by morphine, dibenzyline and adrenal cortex extract. Endocrinology *59:* 347–355 (1956).

Okinaka, S.: Die Regulation der Hypophysen-Nebennierenfunktion durch das Limbic-System und den Mittelhirnanteil der Formatio reticularis. Acta neuroveg. *23:* 15–20 (1961).

Okinaka, S.; Ibayashi, H.; Motohashi, K.; Fujita, T.; Ohsawa, N.; Murakawa, S.: Regulation of the pituitary-adrenocortical function through the limbic system. Neurol. Med. Chir. *2:* 110–115 (1960a).

Okinaka, S.; Ibayashi, H.; Motohashi, K.; Fujita, T.; Yoshida, S.; Ohsawa, N.: Effect of electrical stimulation of the limbic system on pituitary-adrenocortical function: posterior orbital surface. Endocrinology *67:* 319–324 (1960b).

Okinaka, S.; Nakao, K.; Nishikawa, M.; Ikeda, M.; Ibayashi, H.: Studies on the defense mechanism of the body and the neurohumoral regulation. Report III. On the relation between the hormonal secretion of the adrenal cortex and N. splanchnicus. Tohoku J. exp. Med. *56:* 153–159 (1952).

Okuyama, H.; Endo, M.; Ohara, Y.; Takase, S.; Itahara, K.: Circadian rhythm of the plasma cortisol level in cases of prolonged coma. Tohoku J. exp. Med. *123:* 33–47 (1977).

Olling, Ch.C.J.; De Wied, D.: Inhibition of the release of corticotrophin from the hypophysis by chlorpromazine. Acta endocr., Copenh. *22:* 283–292 (1956).

Ondo, J.G.; Mical, R.S.; Porter, J.C.: Passage of radioactive substances from CSF to hypophysial portal blood. Endocrinology *91:* 1239–1246 (1972).

Orth, D.N.; Island, D.P.: Light synchronization of the circadian rhythm in plasma cortisol (17-OHCS) concentration in man. J. clin. Endocr. Metab. *29:* 479–486 (1969).

Orth, D.N.; Island, D.P.; Liddle, G.W.: Experimental alteration of the circadian rhythm in plasma cortisol (17-OHCS) concentration in man. J. clin. Endocr. Metab. *27:* 549–555 (1967).

Osborn, W.; Schoenberg, H.M.; Murphy, J.J.; Erdman, W.J.; Young, D.: Adrenal function in patients with lesions high in the spinal cord. J. Urol. *88:* 1–4 (1962).

Oster, H.L.; Kretchmar, A.L.; Bethell, F.H.: Effect of whole body X-irradiation on ascorbic acid of rat tissues. Proc. Soc. exp. Biol. Med. *84:* 470–473 (1953).

Otsuka, K.: Effects of atropine, eserine and tetramethylammonium on the adrenal 17-hydroxycorticosteroid secretion in anesthetized dogs. Tohoku J. exp. Med. *88:* 165–170 (1966).

Oyama, T.; Saito, T.; Isomatsu, T.; Samejima, N.; Uemura, T.; Arimura, A.: Plasma levels of ACTH and cortisol in man during diethyl ether anesthesia and surgery. Anesthesiology *29:* 559–564 (1968).

Palka, Y.; Coyer, D.; Critchlow, V.: Effects of isolation of medial basal hypothalamus on pituitary-adrenal and pituitary-ovarian functions. Neuroendocrinology *5:* 333–349 (1969).

Papp, M.; Stark, E.; Ács, Zs.; Varga, B.: Kortikoid-Sekretion der Nebenniere bei narkotisierten und wachen Hunden. Endokrinologie *46:* 280–285 (1964).

Patt, H.M.; Swift, M.N.; Tyree, E.B.; John, E.S.: Adrenal response to total body X-radiation. Am. J. Physiol. *150:* 480–487 (1947).

Patt, H.M.; Swift, M.N.; Tyree, E.B.; Straube, R.L.: X irradiation of the hypophysectomized rat. Science *108:* 475–476 (1948).

Pearlmutter, A.F.; Franco-Saenz, R.; Rapino, E.; Saffran, M.: Human adrenal tissue in vitro: steroidogenic properties. J. clin. Endocr. Metab. *39:* 150–153 (1974a).

Pearlmutter, A.F.; Rapino, E.; Saffran, M.: A semi-automated in vitro assay for CRF: activities of peptides related to oxytocin and vasopressin. Neuroendocrinology *15:* 106–119 (1974b).

Peng, T.-C.; Six, K.M.; Munson, P.L.: Effects of prostaglandin E_1 on the hypothalamohypophyseal-adrenocortical axis in rats. Endocrinology *86:* 202–206 (1970).

Perkoff, G.T.; Eik-Nes, K.; Nugent, C.A.; Fred, H.L.; Nimer, R.A.; Rush, L.; Samuels, L.T.; Tyler, F.H.: Studies of the diurnal variation of plasma 17-hydroxycorticosteroids in man. J. clin. Endocr. Metab. *19:* 432–443 (1959).

Péron, F.G.; Moncloa, F.; Dorfman, R.I.: Studies on the possible inhibitory effect of corticosterone on corticosteroidogenesis at the adrenal level in the rat. Endocrinology *67:* 379–388 (1960).

Petersen, P.V.; Weidmann, H.: A study of the effect of various new synthetic compounds on the adrenal ascorbic acid. Acta pharmac. tox. *11*: 103–110 (1955).

Peterson, R.E.: The miscible pool and turnover rate of adrenocortical steroids in man. Recent Prog. Horm. Res. *15*: 231–274 (1959).

Peytremann, A.; Nicholson, W.E.; Hardman, J.G.; Liddle, G.W.: Effect of adrenocorticotropic hormone on extracellular adenosine 3′,5′-monophosphate in the hypophysectomized rat. Endocrinology *92*: 1502–1506 (1973).

Pfeiffer, E.F.; Vaubel, W.E.; Retiene, K.; Berg, D.; Ditschuneit, H.: ACTH-Bestimmung mittels Messung des Plasma-Corticosterons der mit Dexamethason hypophysenblockierten Ratte. Klin. Wschr. *38*: 980–983 (1960).

Pincus, G.: A diurnal rhythm in the excretion of urinary ketosteroids by young men. J. clin. Endocr. Metab. *3*: 195–199 (1943).

Pincus, G.; Hirai, M.: Effects of oestrous cycle variations and exogenous steroid hormones on the production and secretion of corticosterone by the rat adrenal. Acta endocr., Copenh. suppl. *90*: 191–201 (1964).

Plonk, J.W.; Bivens, C.H.; Feldman, J.M.: Inhibition of hypoglycemia-induced cortisol secretion by the serotonin antagonist cyproheptadine. J. clin. Endocr. Metab. *38*: 836–840 (1974).

Plumpton, F.S.; Besser, G.M.: The adrenocortical response to surgery and insulin-induced hypoglycaemia in corticosteroid-treated and normal subjects. Br. J. Surg. *56*: 216–219 (1969).

Pohujani, S.M.; Chittal, S.M.; Raut, V.S.; Sheth, U.K.: Studies in stress induced changes on rat's adrenals. I. Effect of central nervous depressants. Indian J. med. Res. *57*: 1081–1086 (1969).

Poland, R.E.; Weichsel, M.E., Jr.; Rubin, R.T.: Neonatal dexamethasone administration. I. Temporary delay of development of the circadian serum corticosterone rhythm in rats. Endocrinology *108*: 1049–1054 (1981).

Pollock, J.J.; LaBella, F.S.: Inhibition by cortisol of ACTH release from anterior pituitary tissue in vitro. Can. J. Physiol. Pharmacol. *44*: 549–556 (1966).

Popova, N.K.; Maslova, L.N.; Naumenko, E.V.: Serotonin and the regulation of the pituitary-adrenal system after deafferentation of the hypothalamus. Brain Res. *47*: 61–67 (1972).

Portanova, R.; Sayers, G.: Isolated pituitary cells: CRF-like activity of neurohypophysial and related polypeptides. Proc. Soc. exp. Biol. Med. *143*: 661–666 (1973a).

Portanova, R.; Sayers, G.: An in vitro assay for corticotropin releasing factor(s) using suspensions of isolated pituitary cells. Neuroendocrinology *12*: 236–248 (1973b).

Portanova, R.; Sayers, G.: Corticosterone suppression of ACTH secretion: Actinomycin D sensitive and insensitive components of the response. Biochem. biophys. Res. Commun. *56*: 928–933 (1974).

Porter, J.C.: Secretion of corticosterone in rats with anterior hypothalamic lesions. Am. J. Physiol. *204*: 715–718 (1963).

Porter, J.C.: Site of corticotropin-releasing factor (CRF) releasing elements: effect of lesions on ACTH release and adenohypophysial blood flow. Endocrinology *84*: 1398–1403 (1969).

Porter, J.C.; Jones, J.C.: Effect of plasma from hypophyseal-portal vessel blood on adrenal ascorbic acid. Endocrinology *58*: 62–67 (1956).

Porter, J.C.; Klaiber, M.S.: Relationship of input of ACTH to secretion of corticosterone in rats. Am. J. Physiol. *207:*789–792 (1964).

Porter, J.C.; Klaiber, M.S.: Corticosterone secretion in rats as a function of ACTH input and adrenal blood flow. Am. J. Physiol. *209:*811–814 (1965).

Porter, J.C.; Rumsfeld, H.W., Jr.: Effect of lyophilized plasma and plasma fractions from hypophyseal-portal vessel blood on adrenal ascorbic acid. Endocrinology *58:* 359–364 (1956).

Porter, J.C.; Rumsfeld, H.W., Jr.: Further study of an ACTH-releasing protein from hypophyseal portal vessel plasma. Endocrinology *64:*948–954 (1959).

Preziosi, P.; Scapagnini, U.; Nisticó, G.: Brain serotonin depletors and adrenocortical activation. Biochem. Pharmacol. *17:*1309–1313 (1968).

Purves, H.D.; Sirett, N.E.: Corticotrophin secretion by ectopic pituitary glands. Endocrinology *80:*962–968 (1967).

Raisman, G.; Brown-Grant, K.: The 'suprachiasmatic syndrome': endocrine and behavioural abnormalities following lesions of the suprachiasmatic nuclei in the female rat. Proc. R. Soc. Lond. Biol. *198:*297–314 (1977).

Ramaley, J.A.: The effect of an acute light cycle change on adrenal rhythmicity in prepubertal rats. Neuroendocrinology *19:*126–136 (1975).

Ramaley, J.A.; Sieck, G.: Adrenal-gonadal function in rats with frontal hypothalamic transections. Neuroendocrinology *18:*55–64 (1975).

Rastogi, G.K.; Dash, R.J.; Sharma, B.R.; Sawhney, R.C.; Sialy, R.: Circadian responsiveness of the hypothalamic-pituitary axis. J. clin. Endocr. Metab. *42:* 798–803 (1976).

Ratsimamanga, R.: Variations de la teneur en acide ascorbique dans la surrénale au cours du travail. C. r. Séanc. Soc. Biol. *131:*863–865 (1939).

Raymond, L.W.; Sode, J.; Tucci, J.R.: Adrenocortical response to non-exhaustive muscular exercise. Acta endocr., Copenh. *70:*73–80 (1972).

Redgate, E.S.: Spinal cord and adrenocorticotropin release. Proc. Soc. exp. Biol. Med. *105:*528–531 (1960).

Redgate, E.S.; Fahringer, E.E.: A comparison of the pituitary adrenal activity elicited by electrical stimulation of preoptic, amygdaloid and hypothalamic sites in the rat brain. Neuroendocrinology *12:*334–343 (1973).

Redmond, A.F.; Bell, C.C., Jr.: In vitro arterial perfusion of the dog adrenal. Endocrinology *94:*822–828 (1974).

Rerup, C.; Hedner, P.: The effect of pentobarbital (Nembutal, Mebumal NFN) on corticotrophin release in the rat. Acta endocr., Copenh. *39:*518–526 (1962).

Reschini, E.; Giustina, G.: Clinical experience with a simple screening test for Cushing's syndrome combining the determination of plasma cortisol circadian rhythm with the overnight dexamethasone suppression test. Am. J. med. Sci. *275:*33–42 (1978).

Retiene, K.; Schulz, F.: Circadian rhythmicity of hypothalamic CRH and its central nervous regulation. Hormone metabol. Res. *2:*221–224 (1970).

Retiene, K.; Zimmerman, E.; Schindler, W.J.; Neuenschwander, J.; Lipscomb, H.S.: A correlative study of endocrine rhythms in rats. Acta endocr., Copenh. *57:* 615–622 (1968).

Řežábek, K.: The effect of ethyl alcohol on secretion of the adrenocorticotrophic hormone of the pituitary gland in rats. Notes on the mechanism of control of the secretion of ACTH. Physiol. Bohemoslov. *6:*516–522 (1957).

Richards, J.B.: Effects of altered acid-base balance on adrenocortical function in anesthetized dogs. Am. J. Physiol. *188*: 7–11 (1957).

Richards, J.B.; Egdahl, R.H.: Effect of acute hyperthermia on adrenal 17-hydroxycorticosteroid secretion in dogs. Am. J. Physiol. *186*: 435–439 (1956).

Richards, J.B.; Pruitt, R.L.: Hydrocortisone suppression of stress-induced adrenal 17-hydroxycorticosteroid secretion in dogs. Endocrinology *60*: 99–104 (1957).

Richards, J.B.; Stein, S.N.: Effect of carbon dioxide exposure on adrenal 17-hydroxycorticosteroid secretion in dogs. Fed. Proc. *15*: 151 (1956).

Richards, J.B.; Stein, S.N.: Effect of CO_2 exposure and respiratory acidosis on adrenal 17-hydroxycorticosteroid secretion in anesthetized dogs. Am. J. Physiol. *188*: 1–6 (1957).

Riegle, G.D.; Hess, G.D.: Chronic and acute dexamethasone suppression of stress activation of the adrenal cortex in young and aged rats. Neuroendocrinology *9*: 175–187 (1972).

Rijnberk, A.; Der Kinderen, P.J.; Thijssen, J.H.H.: Investigations on the adrenocortical function of normal dogs. J. Endocr. *41*: 387–395 (1968).

Rivas, C.; Borrell, S.: Effects of corticotrophin and dexamethasone on the levels of corticosteroids, adrenaline and noradrenaline in the adrenal glands of cats. J. Endocr. *51*: 283–290 (1971).

Rivkin, I.; Chasin, M.: Nucleotide specificity of the steroidogenic response of rat adrenal cell suspensions prepared by collagenase digestion. Endocrinology *88*: 664–670 (1971).

Rochefort, G.J.; Rosenberger, J.; Saffran, M.: Depletion of pituitary corticotrophin by various stresses and by neurohypophysial preparations. J. Physiol., Lond. *146*: 105–116 (1959).

Ronzoni, E.; Reichlin, S.: Adrenergic agents and the adrenocorticotrophic activity of the anterior pituitary. Am. J. Physiol. *160*: 490–498 (1950).

Rookh, H.V.; Azukizawa, M.; DiStefano, J.J.; Ogihara, T.; Hershman, J.M.: Pituitary-thyroid hormone periodicities in serially sampled plasma of unanesthetized rats. Endocrinology *104*: 851–856 (1979).

Rose, J.C.; Goldsmith, P.C.; Holland, F.J.; Kaplan, S.L.; Ganong, W.F.: Effect of electrical stimulation of the canine brain stem on the secretion of ACTH and growth hormone (GH). Neuroendocrinology *22*: 352–362 (1976).

Rose, R.M.; Gordon, T.P.; Bernstein, I.S.: Diurnal variation in plasma testosterone and cortisol in rhesus monkeys living in social groups. J. Endocr. *76*: 67–74 (1978).

Rose, R.M.; Kreuz, L.E.; Holaday, J.W.; Sulak, K.J.; Johnson, C.E.: Diurnal variation of plasma testosterone and cortisol. J. Endocr. *54*: 177–178 (1972).

Rose, S.; Nelson, J.: Hydrocortisone and A.C.T.H. release. Aust. J. exp. Biol. med. Sci. *34*: 77–80 (1956).

Rosenfeld, R.S.; Rosenberg, B.J.; Fukushima, D.K.; Hellman, L.: 24-Hour secretory pattern of dehydroisoandrosterone and dehydroisoandrosterone sulfate. J. clin. Endocr. Metab. *40*: 850–855 (1975).

Rosenkrantz, H.: A direct influence of 5-hydroxytryptamine on the adrenal cortex. Endocrinology *64*: 355–362 (1959).

Rosenkrantz, H.; Laferte, R.O.: Further observations on the relationship between serotonin and the adrenal. Endocrinology *66*: 832–841 (1960).

Rotsztejn, W.H.; Beaudet, A.; Roberge, A.G.; Lalonde, J.; Fortier, C.: Role of brain serotonin in the circadian rhythm of corticosterone secretion and the corticotropic response to adrenalectomy in the rat. Neuroendocrinology 23: 157-170 (1977).

Roy, B.B.: Effect of anaesthesia on adrenals. Indian med. Forum 20: 91-98 (1969).

Royce, P.C.; Sayers, G.: Blood ACTH: effects of ether, pentobarbital, epinephrine and pain. Endocrinology 63: 794-800 (1958a).

Royce, P.C.; Sayers, G.: Extrapituitary interaction between Pitressin and ACTH. Proc. Soc. exp. Biol. Med. 98: 70-74 (1958b).

Royce, P.C.; Sayers, G.: Biological characterization of a corticotropin releasing factor. Fed. Proc. 18: 132 (1959).

Rubin, R.P.; Carchman, R.A.; Jaanus, S.D.: Role of calcium and adenosine cyclic 3'-5'phosphate in action of adrenocorticotropin. Nature new Biol. 240: 150-152 (1972).

Rubin, R.P.; Warner, W.: Nicotine-induced stimulation of steroidogenesis in adrenocortical cells of the cat. Br. J. Pharmacol. 53: 357-362 (1975).

Rubin, R.T.; Mandell, A.J.; Crandall, P.H.: Corticosteroid responses to limbic stimulation in man: localization of stimulus sites. Science 153: 767-768 (1966).

Ruch, W.; Mixter, R.C.; Russell, R.M.; Garcia, J.F.; Gale, C.C.: Aminergic and thermoregulatory mechanisms in hypothalamic regulation of growth hormone in cats. Am. J. Physiol. 233: E61-E69 (1977).

Russell, S.M.; Dhariwal, A.P.S.; McCann, S.M.; Yates, F.E.: Inhibition by dexamethasone of the in vivo pituitary response to corticotropin-releasing factor (CRF). Endocrinology 85: 512-521 (1969).

Saba, G.C.; Carnicelli, A.; Saba, P.; Marescotti, V.: Diurnal rhythm in the adrenal cortical secretion and in the rate of metabolism of corticosterone in the rat. II. Effects of hypothalamic lesions. Acta endocr., Copenh. 44: 413-415 (1963a).

Saba, G.C.; Hoet, J.J.: The effect of alloxan on the adrenal cortical secretion. I. In male rats. Acta endocr., Copenh. 40: 349-357 (1962).

Saba, G.C.; Saba, P.; Carnicelli, A.; Marescotti, V.: Diurnal rhythm in the adrenal cortical secretion and in the rate of metabolism of corticosterone in the rat. I: In normal animals. Acta endocr., Copenh. 44: 409-412 (1963b).

Saba, P.; Carnicelli, A.; Saba, G.C.; Maltinti, G.; Marescotti, V.: Diurnal rhythm in the adrenal cortical secretion and in the rate of metabolism of corticosterone in the rat. III. In blind animals. Acta endocr., Copenh. 49: 289-292 (1965).

Sadowski, J.; Kurkus, J.; Chwalbińska-Moneta, J.: Plasma hormone and renal function changes in unrestrained dogs exposed to cold. Am. J. Physiol. 228: 376-381 (1975).

Sadowski, J.; Nazar, K.; Szczepańska-Sadowska, E.: Reduced urine concentration in dogs exposed to cold: relation to plasma ADH and 17-OHCS. Am. J. Physiol. 222: 607-610 (1972).

Saffran, M.; Vogt, M.: Depletion of pituitary corticotrophin by reserpine and by a nitrogen mustard. Br. J. Pharmacol. 15: 165-169 (1960).

Sakakura, M.; Saito, Y.; Takebe, K.; Ishii, K.: Studies on fast feedback mechanisms by endogenous glucocorticoids. Endocrinology 98: 954-957 (1976a).

Sakakura, M.; Saito, Y.; Takebe, K.; Yamashita, I.; Ishii, K.: Time course of hypothalamic CRF activity after the administration of two different stresses. Endocr. jap. 23: 413-416 (1976b).

Sakakura, M.; Saito, Y.; Yoshioka, M.; Takebe, K.; Yamashita, I.: Fluctuated changes in the rat hypothalamic content of corticotropin releasing factor around the peak of the circadian rhythm. Tohoku J. exp. Med. *125:* 281–285 (1978).

Sakamaki, T.; Ichikawa, S.; Matsuo, H.: Effect of dexamethasone on the diurnal rhythm of plasma aldosterone in patients with congestive heart failure. Jap. Circul. J. *45:* 739–745 (1981).

Salcman, M.; Peck, L.; Egdahl, R.H.: Effect of acute and prolonged electrical stimulation of the amygdala of the dog upon peripheral plasma concentrations of corticosteroids. Neuroendocrinology *6:* 361–367 (1970).

Sandberg, A.A.; Eik-Nes, K.; Samuels, L.T.; Tyler, F.H.: The effects of surgery on the blood levels and metabolism of 17-hydroxycorticosteroids in man. J. clin. Invest. *33:* 1509–1516 (1954).

Sapeika, N.: The effect of chlorpromazine, iproniazid, and chloroquine on adrenal ascorbic acid in the rat. Archs int. Pharmacodyn. Thér. *122:* 196–200 (1959).

Saruta, T.; Kaplan, N.M.: Adrenocortical steroidogenesis: the effects of prostaglandins. J. clin. Invest. *51:* 2246–2251 (1972).

Sasaki, S.: Studies on the central regulation of the pituitary-adrenocortical system. Acta med. biol. *11:* 185–212 (1963).

Satake, Y.: Secretion of adrenaline and sympathins (Nanzando, Tokyo 1955).

Satake, Y.; Sugawara, T.; Watanabe, M.: A method for collecting the blood from the suprarenal gland in the dog, without fastening, narcotizing, laparotomy or provoking any pain. Tohoku J. exp. Med. *8:* 501–534 (1927).

Sato, T.; Sato, M.; Shinsako, J.; Dallman, M.F.: Corticosterone-induced changes in hypothalamic corticotropin-releasing factor (CRF) content after stress. Endocrinology *97:* 265–274 (1975).

Sayers, G.; Beall, R.J.; Seelig, S.: Isolated adrenal cells: adrenocorticotropic hormone, calcium, steroidogenesis, and cyclic adenosine monophosphate. Science *175:* 1131–1133 (1972).

Sayers, G.; Ma, R.-M.; Giordano, N.D.: Isolated adrenal cells: corticosterone production in response to cyclic AMP (adenosine-3',5'-monophosphate). Proc. Soc. exp. Biol. Med. *136:* 619–622 (1971).

Sayers, G.; Sayers, M.A.: Regulation of pituitary adrenocorticotrophic activity during the response of the rat to acute stress. Endocrinology *40:* 265–273 (1947).

Sayers, G.; Sayers, M.A.; Liang, T.-Y.; Long, C.N.H.: The cholesterol and ascorbic acid content of the adrenal, liver, brain, and plasma following hemorrhage. Endocrinology *37:* 96–110 (1945).

Scapagnini, U.; Annunziato, L.; Di Renzo, G.; Lombardi, G.; Preziosi, P.: Chronic treatment with reserpine and adrenocortical activation. Neuroendocrinology *20:* 243–249 (1976).

Scapagnini, U.; Annunziato, L.; Lombardi, G.; Oliver, Ch.; Preziosi, P.: Time-course of the effect of α-methyl-*p*-tyrosine on ACTH secretion. Neuroendocrinology *18:* 272–276 (1975).

Scapagnini, U.; Moberg, G.P.; Van Loon, G.R.; De Groot, J.; Ganong, W.F.: Relation of brain 5-hydroxytryptamine content to the diurnal variation in plasma corticosterone in the rat. Neuroendocrinology *7:* 90–96 (1971a).

Scapagnini, U.; Van Loon, G.R.; Moberg, G.P.; Ganong, W.F.: Effect of α-methyl-*p*-tyrosine on the circadian variation of plasma corticosterone in rats. Eur. J. Pharmacol. *11:* 266–268 (1970).

Scapagnini, U.; Van Loon, G.R.; Moberg, G.P.; Preziosi, P.; Ganong, W.F.: Evidence for a central adrenergic inhibition of ACTH secretion in rat. Arch. Pharmakol. *269:* 408–409 (1971b).

Scapagnini, U.; Van Loon, G.R.; Moberg, G.P.; Preziosi, P.; Ganong, W.F.: Evidence for central norepinephrine-mediated inhibition of ACTH secretion in the rat. Neuroendocrinology *10:* 155–160 (1972).

Schell, H.; Hornstein, O.P.; Schwarz, W.: Human epidermal cell proliferation with regard to circadian variation of plasma cortisol. Dermatologica *161:* 12–21 (1980).

Scheving, L.E.; Pauly, J.E.: Effect of light on corticosterone levels in plasma of rats. Am. J. Physiol. *210:* 1112–1117 (1966).

Schulster, D.; Tait, S.A.S.; Tait, J.F.; Mrotek, J.: Production of steroids by in vitro superfusion of endocrine tissue. III. Corticosterone output from rat adrenals stimulated by adrenocorticotropin or cyclic 3',5'-adenosine monophosphate and the inhibitory effect of cycloheximide. Endocrinology *86:* 487–502 (1970).

Schwartz, N.B.; Kling, A.: Stress-induced adrenal ascorbic acid depletion in the cat. Endocrinology *66:* 308–310 (1960).

Seelig, S.; Sayers, G.: Bovine hypothalamic corticotropin releasing factor: chemical and biological characteristics. Fed. Proc. *36:* 2100–2103 (1977).

Seggie, J.: Amygdala lesions and 24-h variation in plasma corticosterone, growth hormone, and prolactin levels. Can. J. Physiol. Pharmacol. *58:* 249–253 (1980).

Seggie, J.; Brown, G.M.: Septal lesions and resting adrenal function. A possible explanation of conflicting findings. Neuroendocrinology *8:* 367–374 (1971).

Seggie, J.; Shaw, B.; Uhlir, I.; Brown, G.M.: Baseline 24-hour plasma corticosterone rhythm in normal, sham-operated and septally-lesioned rats. Neuroendocrinology *15:* 51–61 (1974).

Seiden, G.; Brodish, A.: Physiological evidence for 'short-loop' feedback effects of ACTH on hypothalamic CRF. Neuroendocrinology *8:* 154–164 (1971).

Selye, H.; Schenker, V.: A rapid and sensitive method for bioassay of the adrenal cortical hormone. Proc. Soc. exp. Biol. Med. *39:* 518–522 (1938).

Sen, R.N.; Sarangi, N.N.: Effect of lesions in midbrain on ACTH release and control. Indian J. med. Res. *55:* 1072–1077 (1967).

Setchell, K.D.R.; Rees, L.H.; Himsworth, R.L.: Effect of acute inhibition of adrenocorticotrophin secretion on plasma corticosteroids in the rhesus monkey (*Macaca mulatta*). J. Endocr. *67:* 251–257 (1975a).

Setchell, K.D.R.; Shackleton, C.H.L.; Himsworth, R.L.: Studies on plasma corticosteroids in the rhesus monkey (*Macaca mulatta*). J. Endocr. *67:* 241–250 (1975b).

Setekleiv, J.; Skaug, O.E.; Kaada, B.R.: Increase of plasma 17-hydroxycorticosteroids by cerebral cortical and amygdaloid stimulation in the cat. J. Endocr. *22:* 119–127 (1961).

Sevy, R.W.; Ohler, E.A.; Weiner, A.: Effect of chlorpromazine on stress induced adrenal ascorbic acid depletion. Endocrinology *61:* 45–51 (1957).

Shibata, O.: Adrenal cortical and medullary responsiveness to insulin-induced hypoglycemia. Tohoku J. exp. Med. *105:* 27–33 (1971).

Shibata, O.; Narita, S.; Egashira, K.; Waki, S.; Suzuki, T.: Effect of hemorrhage-induced hypotension on adrenocortical responsiveness to adrenocorticotrophin. Tohoku J. exp. Med. *106:* 185–189 (1972).

Shimizu, T.; Mieno, M.; Yamashita, K.: Responses of the hypothalamic-pituitary-adrenal and -gonadal systems to head X-irradiation. Tohoku J. exp. Med. *109:* 155–161 (1973).

Shuster, S.; Williams, I.A.: Pituitary and adrenal function during administration of small doses of corticosteroids. Lancet *ii:* 674–678 (1961).

Siker, E.S.; Lipschitz, E.; Klein, R.: The effect of preanesthetic medications on the blood level of 17-hydroxycorticosteroids. Ann. Surg. *143:* 88–91 (1956).

Silber-Kasprzak, D.; Stępień-Dobrowolska, M.; Chęcińska, E.: 24-hour cortisol profile in essential hypertension. Pol. Tyg. Lek. *30:* 505–506 (1975).

Simon, M.L.; George, R.: Diurnal variations in plasma corticosterone and growth hormone as correlated with regional variations in norepinephrine, dopamine and serotonin content of rat brain. Neuroendocrinology *17:* 125–138 (1975).

Singh, R.K.; Arora, S.R.; Singh, R.C.: Effect of bone injury on adrenocortical circadian periodicity in rabbits. Indian J. exp. Biol. *15:* 1060–1062 (1977).

Singh, R.K.; Chansouria, J.P.N.; Udupa, K.N.: Circadian periodicity of plasma cortisol (17-OHCS) levels in normal, traumatized, corticotrophin and dexamethasone treated rabbits. Indian J. med. Res. *63:* 793–798 (1975).

Sirett, N.E.; Gibbs, F.P.: Dexamethasone suppression of ACTH release: effect of the interval between steroid administration and the application of stimuli known to release ACTH. Endocrinology *85:* 355–359 (1969).

Sirett, N.E.; Purves, H.D.: The assay of corticotrophin-releasing factor in ACTH primed 'grafted' rats; in Brodish, Redgate, Brain-pituitary-adrenal interrelationships, pp. 79–98 (Karger, Basel 1973).

Sithichoke, N.; Marotta, S.F.: Cholinergic influences on hypothalamic-pituitary-adrenocortical activity of stressed rats: an approach utilizing agonists and antagonists. Acta endocr., Copenh. *89:* 726–736 (1978).

Slusher, M.A.: Dissociation of adrenal ascorbic acid and corticosterone responses to stress in rats with hypothalamic lesions. Endocrinology *63:* 412–419 (1958).

Slusher, M.A.: Effect of brainstem lesions on stress-induced corticosteroid release in female rats. Endocrinology *67:* 347–352 (1960).

Slusher, M.A.: Effects of chronic hypothalamic lesions on diurnal and stress corticosteroid levels. Am. J. Physiol. *206:* 1161–1164 (1964).

Slusher, M.A.: Effects of cortisol implants in the brainstem and ventral hippocampus on diurnal corticosteroid levels. Exp. Brain Res. *1:* 184–194 (1966).

Slusher, M.A.; Browning, B.: Morphine inhibition of plasma corticosteroid levels in chronic venous-catheterized rats. Am. J. Physiol. *200:* 1032–1034 (1961).

Slusher, M.A.; Critchlow, V.: Effect of midbrain lesions on ovulation and adrenal response to stress in female rats. Proc. Soc. exp. Biol. Med. *101:* 497–499 (1959).

Slusher, M.A.; Hyde, J.E.: Inhibition of adrenal corticosteroid release by brain stem stimulation in cats. Endocrinology *68:* 773–782 (1961a).

Slusher, M.A.; Hyde, J.E.: Effect of limbic stimulation on release of corticosteroids into the adrenal venous effluent of the cat. Endocrinology *69:* 1080–1084 (1961b).

Slusher, M.A.; Roberts, S.: Effect of hypothalamic lesions on adrenal ascorbic acid response to stress in the male rat. Fed. Proc. *15:* 173 (1956).

Smelik, P.G.: Mechanism of hypophysial response to psychic stress. Acta endocr., Copenh. *33:* 437–443 (1960).

Smelik, P.G.: Failure to inhibit corticotrophin secretion by experimentally induced increases in corticoid levels. Acta endocr., Copenh. *44:* 36–46 (1963).

Smelik, P.G.; Sawyer, C.H.: Effects of implantation of cortisol into the brain stem or pituitary gland on the adrenal response to stress in the rabbit. Acta endocr., Copenh. 41: 561–570 (1962).

Smith, J.J.: The endocrine basis and hormonal therapy of alcoholism. N.Y. St. J. Med. 50: 1704–1706 (1950).

Smith, J.J.: The effect of alcohol on the adrenal ascorbic acid and cholesterol of the rat. J. clin. Endocr. Metab. 11: 792 (1951).

Smith, R.L.; Maickel, R.P.; Brodie, B.B.: ACTH-hypersecretion induced by phenothiazine tranquilizers. J. Pharmac. exp. Ther. 139: 185–190 (1963).

Smolensky, M.H.; Halberg, F.; Harter, J.; Hsi, B.; Nelson, W.: Higher corticosterone values at a fixed single timepoint in serum from mice 'trained' by prior handling. Chronobiologia 5: 1–13 (1978).

Spät, A.; Józan, S.: Effect of prostaglandin E_2 and A_2 on steroid synthesis by the rat adrenal gland. J. Endocr. 65: 55–63 (1975).

Spies, H.G.; Norman, R.L.; Buhl, A.E.: Twenty-four-hour patterns in serum prolactin and cortisol after partial and complete isolation of the hypothalamic-pituitary unit in rhesus monkeys. Endocrinology 105: 1361–1368 (1979).

Staehelin, D.; Labhart, A.; Froesch, R.; Kägi, H.R.: The effect of muscular exercise and hypoglycemia on the plasma level of 17-hydroxysteroids in normal adults and in patients with the adrenogenital syndrome. Acta endocr., Copenh. 18: 521–529 (1955).

Stark, E.; Gyévai, A.; Ács, Zs.; Szalay, K.Sz.; Varga, B.: The site of the blocking action of dexamethasone on ACTH secretion: in vivo and in vitro studies. Neuroendocrinology 3: 275–284 (1968).

Stark, E.; Makara, G.B.; Marton, J.; Palkovits, M.: ACTH release in rats after removal of the medial hypothalamus. Neuroendocrinology 13: 224–233 (1973/74).

Stark, E.; Makara, G.B.; Palkovits, M.; Kárteszi, M.; Mihály, K.: Basal levels of pituitary ACTH and plasma corticosterone after complete or frontal cuts around medial basal hypothalamus. Endocrinol. exp. 12: 209–216 (1978).

Steenburg, R.W.; Smith, L.L.; Moore, F.D.: Conjugated 17-hydroxycorticosteroids in plasma: measurement and significance in relation to surgical trauma. J. clin. Endocr. Metab. 21: 39–52 (1961).

Stern, J.M.; Levine, S.: Psychobiological aspects of lactation in rats. Prog. Brain Res. 41: 433–444 (1974).

Story, J.L.; Melby, J.C.; Egdahl, R.H.; French, L.A.: Adrenal cortical function following stepwise removal of the brain in the dog. Am. J. Physiol. 196: 583–588 (1959).

Studzinski, G.P.; Grant, J.K.: Effect of adenosine-3',5'-phosphate on corticosteroid production in vitro by slices of the adrenal cortex of human beings. Nature, Lond. 193: 1075–1076 (1962).

Sun, C.L.; Thoa, N.B.; Kopin, I.J.: Comparison of the effects of 2-deoxyglucose and immobilization on plasma levels of catecholamines and corticosterone in awake rats. Endocrinology 105: 306–311 (1979).

Sutton, J.R.; Young, J.D.; Lazarus, L.; Hickie, J.B.; Maksvytis, J.: The hormonal response to physical exercise. Aust. Ann. Med. 18: 84–90 (1969).

Suzuki, T.; Abe, K.; Hirose, T.: Adrenal cortical secretion in response to pilocarpine in dogs with hypothalamic lesions. Neuroendocrinology 17: 75–82 (1975a).

Suzuki, T.; Higashi, R.; Hirose, T.; Ikeda, H.; Tamura, K.: Adrenal 17-hydroxycortico-
steroid secretion in the dog in response to ethanol. Acta endocr., Copenh. 70:
736–740 (1972).
Suzuki, T.; Higashi, R.; Tanigawa, H.; Ikeda, H.; Tamura, K.: Adrenal cortical response
to immobilization in conscious and anesthetized dogs. Tohoku J. exp. Med. 94:
281–285 (1968).
Suzuki, T.; Higashi, R.; Tanigawa, H.; Ikeda, H.; Tamura, K.: Effect of ingestion of food
on adrenal 17-hydroxycorticosteroid secretion in the dog. Tohoku J. exp. Med. 105:
303–304 (1971).
Suzuki, T.; Hirai, K.; Otsuka, K.; Matsui, H.; Ohukuzi, S.: Anaphylactic shock and the
secretion of adrenal 17-hydroxycorticosteroid in the dog. Nature, Lond. 211: 1185–
1186 (1966a).
Suzuki, T.; Hirai, K.; Yoshio, H.; Kurouji, K.-I.; Hirose, T.: Effect of eserine and atropine on adrenocortical hormone secretion in unanaesthetized dogs. J. Endocr. 31:
81–82 (1964a).
Suzuki, T.; Hirai, K.; Yoshio, H.; Kurouji, K.; Yamashita, K.: Effect of histamine on adrenal 17-hydroxycorticoid secretion in unanesthetized dogs. Am. J. Physiol. 204:
847–848 (1963).
Suzuki, T.; Hirose, T.; Abe, K.; Matsumoto, I.: Dissociation of adrenocortical secretory
responses to cyanide and pilocarpine in dogs with hypothalamic lesions. Neuroendocrinology 19: 269–276 (1975b).
Suzuki, T.; Ikeda, H.; Narita, S.; Shibata, O.; Waki, S.; Egashira, K.: Adrenal cortical secretion in response to nicotine in conscious and anaesthetized dogs. Q. Jl exp.
Physiol. 58: 139–142 (1973).
Suzuki, T.; Mitamura, T.; Zinnouchi, S.; Yamashita, K.: Der Einfluss des Insulins auf die
17-Hydroxycorticosteroidsekretion der Nebennieren. Naturwissenschaften 51: 219
(1964b).
Suzuki, T.; Ohukuzi, S.; Matsui, H.; Otsuka, K.: Effect of tetramethylammonium and
potassium chloride on adrenal 17-hydroxycorticosteroid secretion in unanesthetized dogs. Tohoku J. exp. Med. 87: 259–261 (1965a).
Suzuki, T.; Otsuka, K.; Matsui, H.; Ohukuzi, S.; Sakai, K.; Harada, Y.: Effect of muscular exercise on adrenal 17-hydroxycorticosteroid secretion in the dog. Endocrinology 80: 1148–1151 (1967).
Suzuki, T.; Romanoff, E.B.; Koella, W.P.; Levy, C.K.: Effect of diencephalic stimuli on
17-hydroxycorticosteroid secretion in unanesthetized dogs. Am. J. Physiol. 198:
1312–1314 (1960).
Suzuki, T.; Sakai, K.; Hirose, T.; Harada, Y.; Higashi, R.; Tanigawa, H.: Adrenal 17-hy-
droxycorticosteroid secretion in response to alloxan in unanaesthetized dogs. Naturwissenschaften 53: 21 (1966b).
Suzuki, T.; Yamashita, K.; Hirai, K.; Kurouji, K.-I.; Yoshio, H.: Adrenal 17-hydroxycor-
ticosteroid secretion in response to cyanide anoxia. Pflügers Arch. ges. Physiol. 285:
119–123 (1965b).
Suzuki, T.; Yamashita, K.; Kamo, M.; Hirai, K.: Effect of sodium pentobarbital and sodium hexobarbital anesthesia on the adrenal 17-hydroxycorticosteroid secretion
rate in the dog. Endocrinology 70: 71–74 (1962).
Suzuki, T.; Yamashita, K.; Mitamura, T.: Muscular exercise and adrenal 17-hydroxycor-
ticosteroid secretion in dogs. Nature, Lond. 181: 715 (1958).

Suzuki, T.; Yamashita, K.; Mitamura, T.: Effect of ether anesthesia on 17-hydroxycorti-costeroid secretion in dogs. Am. J. Physiol. *197:* 1261–1262 (1959a).

Suzuki, T.; Yamashita, K.; Zinnouchi, S.; Mitamura, T.: Effect of morphine upon the adrenal 17-hydroxycorticosteroid secretion in the dog. Nature, Lond. *183:* 825 (1959b).

Suzuki, T.; Yoshio, H.; Kurouji, K.; Hirai, K.: Effect of spinal cord transection on adrenal 17-hydroxycorticosteroid secretion in response to insulin hypoglycaemia. Nature, Lond. *206:* 408 (1965c).

Swan, C.; Abe, K.; Critchlow, V.: Effects of age of blinding on rhythmic pituitary-adrenal function in female rats. Neuroendocrinology *27:* 175–185 (1978).

Szafarczyk, A.; Alonso, G.; Ixart, G.; Malaval, F.; Nouguier-Soule, J.; Assenmacher, I.: Serotoninergic system and circadian rhythms of ACTH and corticosterone in rats. Am. J. Physiol. *239:* E482–E489 (1980a).

Szafarczyk, A.; Hery, M.; Laplante, E.; Ixart, G.; Assenmacher, I.; Kordon, C.: Temporal relationships between the circadian rhythmicity in plasma levels of pituitary hormones and in hypothalamic concentrations of releasing factors. Neuroendocrinology *30:* 369–376 (1980b).

Szafarczyk, A.; Ixart, G.; Alonso, G.; Malaval, F.; Nouguier-Soule, J.; Assenmacher, I.: Effects of raphe lesions on circadian ACTH, corticosterone and motor activity rhythms in free-running blinded rats. Neurosci. Lett. *23:* 87–92 (1981).

Szafarczyk, A.; Ixart, G.; Malaval, F.; Nouguier-Soulé, J.; Assenmacher, I.: Effects of lesions of the suprachiasmatic nuclei and of *p*-chlorophenylalanine on the circadian rhythms of adrenocorticotrophic hormone and corticosterone in the plasma, and on locomotor activity of rats. J. Endocr. *83:* 1–16 (1979).

Szafarczyk, A.; Ixart, G.; Malaval, F.; Nouguier-Soulé, J.; Assenmacher, I.: Corrélation entre les rythmes circadiens de l'ACTH et de la corticostérone plasmatiques, et de l'activité motrice, évoluant en «libre cours» après énucléation oculaire chez le rat. C. r. hebd. Séanc. Acad. Sci., Paris sér. D *290:* 587–592 (1980c).

Tait, S.A.S.; Tait, J.F.; Okamoto, M.; Flood, C.: Production of steroids by in vitro superfusion of endocrine tissue. I. Apparatus and a suitable analytical method for adrenal steroid output. Endocrinology *81:* 1213–1225 (1967).

Takahashi, K.; Hanada, K.; Kobayashi, K.; Hayafuji, C.; Otani, S.; Takahashi, Y.: Development of the circadian adrenocortical rhythm in rats: studied by determination of 24- or 48-hour patterns of blood corticosterone levels in individual pups. Endocrinology *104:* 954–961 (1979).

Takahashi, K.; Inoue, K.; Kobayashi, K.; Hayafuji, C.; Nakamura, Y.; Takahashi, Y.: Effects of food restriction on circadian adrenocortical rhythm in rats under constant lighting conditions. Neuroendocrinology *23:* 193–199 (1977a).

Takahashi, K.; Inoue, K.; Takahashi, Y.: No effect of pinealectomy on the parallel shift in circadian rhythms of adrenocortical activity and food intake in blinded rats. Endocr. jap. *23:* 417–421 (1976).

Takahashi, K.; Inoue, K.; Takahashi, Y.: Parallel shift in circadian rhythms of adrenocortical activity and food intake in blinded and intact rats exposed to continuous illumination. Endocrinology *100:* 1097–1107 (1977b).

Takahashi, Y.; Kipnis, D.M.; Daughaday, W.H.: Growth hormone secretion during sleep. J. clin. Invest. *47:* 2079–2090 (1968).

Takebe, K.; Kunita, H.; Sakakura, M.; Horiuchi, Y.: Effect of dexamethasone on ACTH release induced by lysine vasopressin in man; time interval between dexamethasone and vasopressin injection. J. clin. Endocr. Metab. *28:* 644–650 (1968).

Takebe, K.; Kunita, H.; Sakakura, M.; Horiuchi, Y.; Mashimo, K.: Suppressive effect of dexamethasone on the rise of CRF activity in the median eminence induced by stress. Endocrinology *89:* 1014–1019 (1971a).

Takebe, K.; Sakakura, M.: Circadian rhythm of CRF activity in the hypothalamus after stress. Endocr. jap. *19:* 567–570 (1972).

Takebe, K.; Sakakura, M.; Brodish, A.: Studies on long feedback via glucocorticoid and short feedback via ACTH; in Hatotani, Psychoneuroendocrinology, Workshop Conf. Int. Soc. Psychoneuroendocrinol., Mieken 1973, pp. 198–205 (Karger, Basel 1974).

Takebe, K.; Sakakura, M.; Horiuchi, Y.; Mashimo, K.: Persistence of diurnal periodicity of CRF activity in adrenalectomized and hypophysectomized rats. Endocr. jap. *18:* 451–455 (1971b).

Takebe, K.; Sakakura, M.; Mashimo, K.: Continuance of diurnal rhythmicity of CRF activity in hypophysectomized rats. Endocrinology *90:* 1515–1520 (1972).

Takebe, K.; Setaishi, C.; Hirama, M.; Yamamoto, M.; Horiuchi, Y.: Effects of a bacterial pyrogen on the pituitary-adrenal axis at various times in the 24 hours. J. clin. Endocr. Metab. *26:* 437–442 (1966).

Takeuchi, A.; Kajihara, A.; Suzuki, M.: Effect of acute exposure to cold on the levels of corticosterone and pituitary hormones in plasma collected from free conscious cannulated rats. Endocr. jap. *24:* 109–114 (1977).

Tang, F.; Phillips, J.G.: Pituitary-adrenal response in male rats subjected to continuous ether stress and the effect of dexamethasone treatment. J. Endocr. *75:* 325–326 (1977).

Tang, P.C.; Patton, H.D.: Effect of hypophysial stalk section on adenohypophysial function. Endocrinology *49:* 86–98 (1951).

Tanigawa, H.: On the mechanism of stimulatory action of histamine upon the adrenal 17-hydroxycorticosteroid secretion in the dog. Tohoku J. exp. Med. *92:* 281–289 (1967).

Tanigawa, H.; Higashi, R.; Suzuki, T.: Effect of hypophysectomy on the adrenal 17-hydroxycorticosteroid secretion in response to eserine in dogs. Tohoku J. exp. Med. *94:* 393–396 (1968).

Tavadia, H.B.; Fleming, K.A.; Hume, P.D.; Simpson, H.W.: Circadian rhythmicity of human plasma cortisol and PHA-induced lymphocyte transformation. Clin. exp. Immunol. *22:* 190–193 (1975).

Taylor, A.N.; Branch, B.J.: Interactions of forebrain and bulbar inhibitory mechanisms with the reticular activating system in the control of ACTH release. Fed. Proc. *28:* 438 (1969).

Taylor, A.N.; Farrell, G.: Effects of brain stem lesions on aldosterone and cortisol secretion. Endocrinology *70:* 556–566 (1962).

Tepperman, J.; Rakieten, N.; Birnie, J.H.; Diermeier, H.F.: Effect of antihistamine drugs on the adrenal cortical response to histamine and to stress. J. Pharmac. exp. Ther. *101:* 144–152 (1951).

Terpstra, J.; Hessel, L.W.; Seepers, J.; Van Gent, C.M.: The influence of meal frequency on diurnal lipid, glucose and cortisol levels in normal subjects. Eur. J. clin. Invest. *8:* 61–66 (1978).

Tilders, F.J.H.; Berkenbosch, F.; Smelik, P.G.: Adrenergic mechanisms involved in the control of pituitary-adrenal activity in the rat: a β-adrenergic stimulatory mechanism. Endocrinology *110:* 114–120 (1982).

Toft, H.; Buus, O.; Nielsen, E.: Vasopressin in the diagnostic evaluation of pituitary and hypothalamic function. Acta endocr., Copenh. *67:* 393–400 (1971).

Tronchetti, F.; Marescotti, V.; Saba, G.C.; Carnicelli, A.; Saba, P.: Investigations on control of adrenocortical activity by the nervous system. Eur. Rev. Endocrinol. *1:* 281–290 (1965).

Tucci, J.R.; Espiner, E.A.; Jagger, P.I.; Lauler, D.P.; Thorn, G.W.: Vasopressin in the evaluation of pituitary-adrenal function. Ann. intern. Med. *69:* 191–202 (1968).

Tullner, W.W.; Hertz, R.: Suppression of corticosteroid production in the dog by Monase. Proc. Soc. exp. Biol. Med. *116:* 837–840 (1964).

Turton, M.B.; Deegan, T.: Circadian variations of plasma catecholamine, cortisol and immunoreactive insulin concentrations in supine subjects. Clinica chim. Acta *55:* 389–397 (1974).

Tyler, F.H.; Schmidt, C.D.; Eik-Nes, K.; Brown, H.; Samuels, L.T.: The role of the liver and the adrenal in producing elevated plasma 17-hydroxycorticosteroid levels in surgery. J. clin. Invest. *33:* 1517–1523 (1954).

Uhlir, I.; Seggie, J.; Brown, G.M.: The effect of septal lesions on the threshold of adrenal stress response. Neuroendocrinology *14:* 351–355 (1974).

Ulrich, R.; Yuwiler, A.; Geller, E.: Neonatal hydrocortisone: effect on the development of the stress response and the diurnal rhythm of corticosterone. Neuroendocrinology *21:* 49–57 (1976).

Ulrich, R.S.; Yuwiler, A.: Failure of 6-hydroxydopamine to abolish the circadian rhythm of serum corticosterone. Endocrinology *92:* 611–614 (1973).

Urquhart, J.: Adrenal blood flow and the adrenocortical response to corticotropin. Am. J. Physiol. *209:* 1162–1168 (1965).

Usher, D.R.; Kasper, P.; Birmingham, M.K.: Comparison of pituitary-adrenal function in rats lesioned in different areas of the limbic system and hypothalamus. Neuroendocrinology *2:* 157–174 (1967).

Van Cantfort, J.: Influence de la période d'éclairement et du moment de prise de nourriture dans le contrôle de l'activité circadienne de la cholestérol-7α-hydroxylase chez le rat. C. r. hebd. Séanc. Acad. Sci., Paris sér. D *279:* 1273–1276 (1974).

Van Cauter, E.; Golstein, J.; Vanhaelst, L.; Leclercq, R.: Effects of oral contraceptive therapy on the circadian patterns of cortisol and thyrotropin (TSH). Eur. J. clin. Invest. *5:* 115–121 (1975).

Van Cauwenberge, H.: Relation of salicylate action to pituitary gland. Observations in rats. Lancet *ii:* 374–375 (1951).

Van Cauwenberge, H.: Contribution à l'étude de la réactivité surrénalienne du rat. Archs int. Pharmacodyn. Thér. *106:* 473–599 (1956).

Van Cauwenberge, H.; Fischer, P.; Vliers, M.; Bacq, Z.M.: Influence d'une irradiation létale sur le taux des stéroïdes réducteurs sanguins chez le rat. C. r. Séanc. Soc. Biol. *151:* 198–201 (1957a).

Van Cauwenberge, H.; Fischer, P.; Vliers, M.; Bacq, Z.M.: Irradiation du rat et taux des 17-hydroxycorticostéroïdes sanguins. Archs int. Physiol. Biochim. *65:* 143–147 (1957b).

Van Delft, A.M.L.; Kaplanski, J.; Smelik, P.G.: Circadian periodicity of pituitary-adrenal function after *p*-chlorophenylalanine administration in the rat. J. Endocr. *59:* 465–474 (1973).

Van der Wal, B.; Israëls, A.L.M.; Janssen, J.F.; De Wied, D.: Some clinical experience in the determination of cortisol and corticosterone in a small sample of human plasma. Acta endocr., Copenh. *38:* 392–398 (1961).

Van Loon, G.R.; Hilger, L.; King, A.B.; Boryczka, A.T.; Ganong, W.F.: Inhibitory effect of *L*-dihydroxyphenylalanine on the adrenal venous 17-hydroxycorticosteroid response to surgical stress in dogs. Endocrinology *88:* 1404–1414 (1971a).

Van Loon G.R.; Scapagnini, U.; Cohen, R.; Ganong, W.F.: Effect of intraventricular administration of adrenergic drugs on the adrenal venous 17-hydroxycorticosteroid response to surgical stress in the dog. Neuroendocrinology *8:* 257–272 (1971b).

Van Loon, G.R.; Scapagnini, U.; Moberg, G.P.; Ganong, W.F.: Evidence for central adrenergic neural inhibition of ACTH secretion in the rat. Endocrinology *89:* 1464–1469 (1971c).

Van Peenen, P.F.D.; Way, E.L.: The effect of certain central nervous system depressants on pituitary-adrenal activating agents. J. Pharmac. exp. Ther. *120:* 261–267 (1957).

Van Ree, J.M.; Spaapen-Kok, W.B.; De Wied, D.: Differential localization of pituitary-adrenal activation and temperature changes following intrahypothalamic microinjection of morphine in rats. Neuroendocrinology *22:* 318–324 (1976a).

Van Ree, J.M.; Versteeg, D.H.G.; Spaapen-Kok, W.B.; De Wied, D.: Effects of morphine on hypothalamic noradrenaline and on pituitary-adrenal activity in rats. Neuroendocrinology *22:* 305–317 (1976b).

Vapaatalo, H.; Bieck, P.; Westermann, E.: Actions of various cyclic nucleotides, nucleosides and purine bases on the synthesis of corticosterone in vitro. Arch. Pharmacol. *275:* 435–443 (1972).

Venters, K.D.; Painter, E.E.: Pituitary and adrenal ascorbic acid in x-irradiated rats. Fed. Proc. *10:* 141 (1951).

Verdesca, A.S.; Westermann, C.D.; Crampton, R.S.; Black, W.C.; Nedeljkovic, R.I.; Hilton, J.G.: Direct adrenocortical stimulatory effect of serotonin. Am. J. Physiol. *201:* 1065–1067 (1961).

Vermes, I.; Mulder, G.H.; Smelik, P.G.: A superfusion system technique for the study of the sites of action of glucocorticoids in the rat hypothalamus-pituitary-adrenal system in vitro. II. Hypothalamus-pituitary cell-adrenal cell superfusion. Endocrinology *100:* 1153–1159 (1977).

Vermes, I.; Telegdy, G.: Effect of intraventricular injection and intrahypothalamic implantation of serotonin on the hypothalamo-hypophyseal-adrenal system in the rat. Acta physiol. hung. *42:* 49–59 (1972).

Vermes, I.; Telegdy, G.; Lissák, K.: Inhibitory action of serotonin on hypothalamus-induced ACTH release. Acta physiol. hung. *41:* 95–98 (1972).

Vernikos, J.; Dallman, M.F.; Bonner, C.; Katzen, A.; Shinsako, J.: Pituitary-adrenal function in rats chronically exposed to cold. Endocrinology *110:* 413–420 (1982).

Vernikos-Danellis, J.: Estimation of corticotropin-releasing activity of rat hypothalamus and neurohypophysis before and after stress. Endocrinology *75:* 514–520 (1964).

Vernikos-Danellis, J.: Effect of stress, adrenalectomy, hypophysectomy and hydrocortisone on the corticotropin-releasing activity of rat median eminence. Endocrinology *76:* 122–126 (1965).

Vernikos-Danellis, J.; Berger, P.; Barchas, J.D.: Brain serotonin and pituitary-adrenal function. Prog. Brain Res. *39*: 301–310 (1973).

Vernikos-Danellis, J.; Kellar, K.J.; Kent, D.; Gonzales, C.; Berger, P.A.; Barchas, J.D.: Serotonin involvement in pituitary-adrenal function. Ann. N.Y. Acad. Sci. *297*: 518–526 (1977).

Vernikos-Danellis, J.; Marks, B.H.: Pituitary inhibitory effects of digitoxin and hydrocortisone. Proc. Soc. exp. Biol. Med. *109*: 10–14 (1962).

Vinson, G.P.: The relationship between corticosterone synthesis from endogenous precursors and from added radioactive precursors by rat adrenal tissue in vitro. J. Endocr. *36*: 231–238 (1966).

Virtue, R.W.; Helmreich, M.L.; Gainza, E.: The adrenal cortical response to surgery. I. The effect of anesthesia on plasma 17-hydroxycorticosteroid levels. Surgery, St. Louis *41*: 549–566 (1957).

Viru, A.A.: Role of the adrenocortical response to physical exertion in increased working capacity of athletes. Bull. Exp. Biol. Med. *82*: 957–959 (1976).

Viru, A.; Äkke, H.: Effects of muscular work on cortisol and corticosterone content in the blood and adrenals of guinea pigs. Acta endocr., Copenh. *62*: 385–390 (1969).

Vogt, M.: The output of cortical hormone by the mammalian suprarenal. J. Physiol., Lond. *102*: 341–356 (1943).

Vogt, M.: Observations on some conditions affecting the rate of hormone output by the suprarenal cortex. J. Physiol., Lond. *103*: 317–332 (1944).

Vogt, M.: Cortical secretion of the isolated perfused adrenal. J. Physiol., Lond. *113*: 129–156 (1951a).

Vogt, M.: The role of hypoglycaemia and of adrenaline in the response of the adrenal cortex to insulin. J. Physiol., Lond. *114*: 222–233 (1951b).

Voloschin, L.; Joseph, S.A.; Knigge, K.M.: Endocrine function in male rats following complete and partial isolations of the hypothalamo-pituitary unit. Neuroendocrinology *3*: 387–397 (1968).

Vukoson, M.B.; Kramer, R.E.; Pope, M.; Greiner, J.W.; Colby, H.D.: Failure of indomethacin to affect adrenal responsiveness to ACTH in vitro. Hormone metabol. Res. *8*: 325–326 (1976).

Walker, W.F.; Shoemaker, W.C.; Kaalstad, A.J.; Moore, F.D.: Influence of blood volume restoration and tissue trauma on corticosteroid secretion in dogs. Am. J. Physiol. *197*: 781–785 (1959a).

Walker, W.F.; Zileli, M.S.; Reutter, F.W.; Shoemaker, W.C.; Moore, F.D.: Factors influencing the 'resting' secretion of the adrenal medulla. Am. J. Physiol. *197*: 765–772 (1959b).

Walter-Van Cauter, E.; Virasoro, E.; Leclercq, R.; Copinschi, G.: Seasonal, circadian and episodic variations of human immunoreactive β-MSH, ACTH and cortisol. Int. J. Pept. Prot. Res. *17*: 3–13 (1981).

Warner, W.; Rubin, R.P.: Evidence for a possible prostaglandin link in ACTH-induced steroidogenesis. Prostaglandins *9*: 83–95 (1975).

Way, E.L.; Taylor, J.; Old, L.J.; George, R.: Corticotrophic studies with aspirin. J. Endocr. *24*: 7–16 (1962).

Weidmann, H.: Ascorbinsäureherabsetzung in den Nebennieren hypophysektomierter Ratten durch Salicylsäure. Arch. exp. Path. Pharmak. *225*: 342–345 (1955).

Weinges, K.F.; Schwarz, K.: Der Einfluss von Glucagon, Insulin und Glucose auf die freien 17-Hydroxycorticosteroide im Plasma beim Menschen. Klin. Wschr. *38:* 792–796 (1960).

Weitzman, E.D.; Fukushima, D.; Nogeire, C.; Roffwarg, H.; Gallagher, T.F.; Hellman, L.: Twenty-four hour pattern of the episodic secretion of cortisol in normal subjects. J. clin. Endocr. Metab. *33:* 14–22 (1971).

Weitzman, E.D.; Schaumburg, H.; Fishbein, W.: Plasma 17-hydroxycorticosteroid levels during sleep in man. J. clin. Endocr. Metab. *26:* 121–127 (1966).

Wells, H.; Briggs, F.N.; Munson, P.L.: The inhibitory effect of reserpine on ACTH secretion in response to stressful stimuli. Endocrinology *59:* 571–579 (1956).

Wenzkat, P.B.; Hubl, W.; Kirsch, K.R.; Büchner, M.: Die Wirkung von Muskelarbeit auf den Harn- und Plasmaspiegel der unkonjugierten 11-Hydroxykortikosteroide. Acta biol. med. germ. *21:* 163–168 (1968).

Werk, E.E., Jr.; Garber, S.; Sholiton, L.J.: Effect of sympathetic blockade on changes in blood ketones and nonesterified fatty acids following hypoglycemia in man. Metabolism *10:* 115–125 (1961).

Westermann, E.O.; Maickel, R.P.; Brodie, B.B.: On the mechanism of pituitary-adrenal stimulation by reserpine. J. Pharmac. exp. Ther. *138:* 208–217 (1962).

Wexler, B.C.: Effects of a bacterial polysaccharide (Piromen) on the pituitary-adrenal axis: modification of ACTH release by morphine and salicylate. Metabolism *12:* 49–56 (1963).

Wexler, B.C.; Dolgin, A.E.; Tryczynski, E.W.: Effects of a bacterial polysaccharide (Piromen) on the pituitary-adrenal axis: adrenal ascorbic acid, cholesterol and histologic alterations. Endocrinology *61:* 300–308 (1957).

Wexler, B.C.; Dolgin, A.E.; Zaroslinski, J.F.; Tryczynski, E.W.: Effects of a bacterial polysaccharide (Piromen) on the pituitary-adrenal axis: cortisone blockade of Piromen-induced release of ACTH. Endocrinology *63:* 201–204 (1958).

Wexler, B.C.; Pencharz, R.; Thomas, S.F.: Adrenal ascorbic acid in male and female rats after total body X-ray irradiation. Proc. Soc. exp. Biol. Med. *79:* 183–184 (1952).

Wexler, B.C.; Pencharz, R.; Thomas, S.F.: Adrenal ascorbic acid and histological changes in male and female rats after half-body X-ray irradiation. Am. J. Physiol. *183:* 71–74 (1955).

Wilcox, C.S.; Aminoff, M.J.; Millar, J.G.B.; Keenan, J.; Kremer, M.: Circulating levels of corticotrophin and cortisol after infusions of *L*-dopa, dopamine and noradrenaline, in man. Clin. Endocrinol. *4:* 191–198 (1975).

Wiley, M.K.; Pearlmutter, A.F.; Miller, R.E.: Decreased adrenal sensitivity to ACTH in the vasopressin-deficient (Brattleboro) rat. Neuroendocrinology *14:* 257–270 (1974).

Wilkinson, C.W.; Shinsako, J.; Dallman, M.F.: Daily rhythms in adrenal responsiveness to adrenocorticotropin are determined primarily by the time of feeding in the rat. Endocrinology *104:* 350–359 (1979).

Wilson, M.; Critchlow, V.: Effect of fornix transection or hippocampectomy on rhythmic pituitary-adrenal function in the rat. Neuroendocrinology *13:* 29–40 (1973/74).

Wilson, M.; Critchlow, V.: Effect of septal ablation on rhythmic pituitary-adrenal function in the rat. Neuroendocrinology *14:* 333–344 (1974).

Wilson, M.I.; Brown, G.M.; Wilson, D.: Annual and diurnal changes in plasma androgen and cortisol in adult male squirrel monkeys (*Saimiri sciureus*) studied longitudinally. Acta endocr., Copenh. *87:* 424–433 (1978).

Wilson, M.M.; Critchlow, V.: Absence of a circadian rhythm in persisting corticosterone fluctuations following surgical isolation of the medial basal hypothalamus. Neuroendocrinology *19:* 185–192 (1975).

Wilson, M.M.; Greer, M.A.: Evidence for a free-running pituitary-adrenal circadian rhythm in constant light-treated adult rats. Proc. Soc. exp. Biol. Med. *154:* 69–71 (1977).

Wilson, M.M.; Rice, R.W.; Critchlow, V.: Evidence for a free-running circadian rhythm in pituitary-adrenal function in blinded adult female rats. Neuroendocrinology *20:* 289–295 (1976).

Wise, B.L.; Pont, M.; Ganong, W.F.: Failure of hind-brain removal to depress ACTH secretion in dogs with isolated pituitaries. Fed. Proc. *21:* 196 (1962).

Wise, B.L.; Van Brunt, E.E.; Ganong, W.F.: Effect of removal of various parts of the brain on ACTH secretion in dogs. Proc. Soc. exp. Biol. Med. *112:* 792–795 (1963).

Witorsch, R.J.; Brodish, A.: Conditions for the reliable use of lesioned rats for the assay of CRF in tissue extracts. Endocrinology *90:* 552–557 (1972).

Wolf, R.C.; Bowman, R.E.: Adrenal function in the rhesus monkey after total-body X-irradiation. Radiat. Res. *23:* 232–238 (1964).

Yamaguchi, N.; Maeda, K.; Kuromaru, S.: The effects of sleep deprivation on the circadian rhythm of plasma cortisol levels in depressive patients. Folia psychiat. neurol. jap. *32:* 479–487 (1978).

Yamashita, K.; Shimizu, T.: Responsiveness of the X-irradiated adrenal cortex to adrenocorticotrophin in dogs. J. Endocr. *52:* 199–200 (1972).

Yamashita, K.; Shimizu, T.; Mieno, M.; Kawao, K.: The susceptibility of the hypothalamic-pituitary-adrenocortical axis to histamine after head X-irradiation. Tohoku J. exp. Med. *110:* 161–166 (1973).

Yamashita, K.; Shimizu, T.; Mieno, M.; Yamashita, E.: Effects of exogenous acetylcholine upon adrenal 17-hydroxycorticosteroid secretion of intact and head X-irradiated dogs. Neuroendocrinology *27:* 39–45 (1978).

Yasuda, N.; Greer, M.A.: Studies on the corticotrophin-releasing activity of vasopressin, using ACTH secretion by cultured rat adenohypophyseal cells. Endocrinology *98:* 936–942 (1976a).

Yasuda, N.; Greer, M.A.: Rat hypothalamic corticotropin-releasing factor (CRF) content remains constant despite marked acute or chronic changes in ACTH secretion. Neuroendocrinology *22:* 48–56 (1976b).

Yasuda, N.; Greer, M.A.: Evidence that the hypothalamus mediates endotoxin stimulation of adrenocorticotropic hormone secretion. Endocrinology *102:* 947–953 (1978).

Yasuda, N.; Greer, M.A.; Greer, S.E.; Panton, P.: Studies on the site of action of vasopressin in inducing adrenocorticotropin secretion. Endocrinology *103:* 906–911 (1978).

Yasuda, N.; Takebe, K.; Greer, M.A.: Evidence of nycterohemeral periodicity in stress-induced pituitary-adrenal activation. Neuroendocrinology *21:* 214–224 (1976).

Yates, F.E.; Leeman, S.E.; Glenister, D.W.; Dallman, M.F.: Interaction between plasma corticosterone concentration and adrenocorticotropin-releasing stimuli in the rat: evidence for the reset of an endocrine feedback control. Endocrinology *69:* 67–80 (1961).

Yates, F.E.; Russell, S.M.; Dallman, M.F.; Hedge, G.A.; McCann, S.M.; Dhariwal,

A.P.S.: Potentiation by vasopressin of corticotropin release induced by corticotropin-releasing factor. Endocrinology 88:3–15 (1971).

Yoshio, H.: Inhibitory effect of pre-optic stimulation on adrenal 17-hydroxycorticosteroid secretion rate in the cat. Nature, Lond. 201:1334–1335 (1964).

Young, A.K.; Walker, B.L.: The circadian cycles of plasma corticosterone and adrenal cholesteryl esters in the normal and EFA-deficient female rat. Lipids 13: 181–186 (1978).

Ziegler, D.K.; Hassanein, R.S.; Kodanaz, A.; Meek, J.C.: Circadian rhythms of plasma cortisol in migraine. J. Neurol. Neurosurg. Psychiat. 42:741–748 (1979).

Zimmermann, E.; Critchlow, V.: Effects of diurnal variation in plasma corticosterone levels on adrenocortical response to stress. Proc. Soc. exp. Biol. Med. 125:658–663 (1967).

Zimmermann, E.; Critchlow, V.: Suppression of pituitary-adrenal function with physiological plasma levels of corticosterone. Neuroendocrinology 5:183–192 (1969a).

Zimmermann, E.; Critchlow, V.: Negative feedback and pituitary-adrenal function in female rats. Am. J. Physiol. 216:148–155 (1969b).

Zimmermann, E.; Critchlow, V.: Effects of intracerebral dexamethasone on pituitary-adrenal function in female rats. Am. J. Physiol. 217:392–396 (1969c).

Zimmermann, E.; Smyrl, R.; Critchlow, V.: Suppression of pituitary-adrenal response to stress with physiological plasma levels of corticosterone in the female rat. Neuroendocrinology 10:246–256 (1972).

Zukoski, C.F.: Correlation between adrenal and gastric secretion induced by insulin hypoglycemia. Sth. med. J., Nashville 57:1440–1442 (1964).

Zukoski, C.F.: Mechanism of action of insulin hypoglycemia on adrenal cortical secretion. Endocrinology 78:1264–1267 (1966).

Zukoski, C.F.; Ney, R.L.: ACTH secretion after pituitary stalk section in the dog. Am. J. Physiol. 211:851–854 (1966).

Author Index

Subject Index